# Short Stories of Underage American Military Warriors

Short stories of young men who served in the Armed Forces of the United States of America before attaining legal age.

Compiled and edited by
Ophelias "Joe" Bourque and
Yvonne "Bonnie" Bourque

# Short Stories of Underage American Military Warriors



Additional copies of this book available from the publisher:

Joe and Bonnie Bourque
P.O. Box 1072
Jennings, LA 70546

Many of the excerpts therein are reprinted from
*America's Young Warriors*



Printing in the USA by
Wise Publications
809 East Napoleon
Sulphur, Louisiana
337-527-8308

# Contents

# A FEW NOTES TO THE READER

The organization known as the Veterans of Underage Military Service was formed in 1991 by Allan Stover. At the time of this writing, we have a total of 1,302 members from throughout the United States. Mr. Stover had begun work on a book listing the names and information about the members but because of a transfer with his employment, the work was not completed by him.

In 1995, Ray Jackson and his wife, Susan, began the project to complete this work. This required the task of editing and arranging the stories sent to them by VUMS members. The first edition was finished in 1996. Eventually, two additional volumes followed. The Jacksons labored on this monumental task for a period of twelve years. They are to be commended for their diligence and dedication.

My wife, Bonnie, and I attended the VUMS National Convention in April of 2009 as it convened in the City of St. Louis, Missouri. We had a wonderful time and were able to meet a large number of underage veterans. At this time, there was much discussion concerning the compilation and publication of a fourth volume of records for those who were not included in volumes one, two and three.

I spoke with Mr. Jackson concerning this. He informed me that due to his health, he would not be able to undertake the work on another volume. At this point, I informed the VUMS members that I would undertake this project.

Due to mitigating circumstances, the VUMS leaders informed me that the organization would not be able to underwrite the expenses incurred in the completion of another volume. As I reflected on the importance of such a work, I made the commitment to proceed with the task at my own expense.

With the participation of many people providing the stories, my wife and I have been able to bring the book to birth. This volume is entitled Short Stories of America's Underage Warriors, so as not to infringe upon the copyright of the first three volumes.

The stories contained in this volume reflect the motivation and interest that the individuals listed had in their hearts when as teens who were under the required age, they chose to enter the Military. A number of these men were involved in combat during World War Two, the Korean War, and the Vietnam War, so the present age range of these men is quite wide. Some of them rose through the ranks from Private all the way to General, from Seaman all the way to Admiral.

One of the factors that has made The United States of America the great country that it is, is to be found in men and women such as those whose deployment led them to make the commitment and sacrifices that they did. It is by such as these who all down through history were willing even to give their very lives so that others may live in freedom and at peace.

*Ophelias "Joe" Bourque*

Ophelias “Joe” Bourque
and
Yvonne “Bonnie” Bourque

*(As this volume went to press in April 2010, Joe Bourque began receiving treatment at M.D. Anderson Cancer Center in Houston. He has been diagnosed with Non-Hodgkins Lymphoma, Acute Lymphcytic Leukemia, and Bone Cancer. Progress is good with his loving wife Bonnie everalways at his side.*

# ACKNOWLEDGEMENTS

Once we began our work on this book we knew we would be needing help and guidance to bring this project to a conclusion. We would like to express our appreciation to the following people for their contributions, their time and talents.

I would like to thank each one of the members who have contributed their life story to be entered into this volume. Bonnie and I would like for you to know the respect we have for each one of you. Your stories have been recorded as historical facts and we feel that your families will cherish these accounts for years to come. We are reminded as to the need for our United States Government to continue showing respect for and assistance to our veterans.

We would like to thank William "Bill" Morgan, who serves the VUMS organization as the National Vice Commander. He was instrumental in obtaining some of the VUMS stories for us. The time spent on telephone calls as well as all of the other contributions he has made to help Bonnie and I to be able to complete this book.

Mr. Ray Jackson has been of great assistance to me. I inquired of him concerning permission to use some inserts from one of his books. He readily agreed, providing proper credit would be given.

Our minister, Bro. Jack D. Harris contributed the poetic works in which evidence is shown concerning how war changes individuals.

My wife, Yvonne, whom I call Bonnie, agreed to help me with this project, which has often been quite worrisome. Our computer has consistently proven to be troublesome, especially in the bad weather conditions, etc. We have had printing problems and have suffered the loss of some of the things we desired to keep. On two different occasions, she had to enter the hospital, but through it all, she never lost sight of what we were trying to accomplish. I deeply appreciate her help and the fact that she has shared this editorship with me.

# I Gave My Life For Your Freedom

*By Ophelias "Joe" Bourque*

Many years ago, I reached the shores of a land that eventually would be known as the United States of America. I had left the lands on the eastern side of the Atlantic Ocean in search of a location in which I could grow and develop without the oppression I had known earlier. After arriving in this new land, it was not long until I realized that in order to remain here and pursue my dreams, it would require much sacrifice.

I soon learned that there were natives living in this country that would resent my entry and would try to oppose any advancement that I may desire to make. In addition to this threat, there was the mighty force of the British army that had been sent by the King of England to try and subject me to the crown.

When I was confronted with this hostility, under General George Washington's command, we fought bravely and with determination. At Bunker Hill, there was a bloody engagement with this enemy. I fought long and hard but during the encounter, I suffered a bullet wound in my left side. As long as I could stand on my feet, I kept fighting, but due to a terrific loss of blood, I fell on the field of battle. As I lay dying, I knew in my heart that it was worthwhile. I gave my life so others may live in freedom. At the time of my death, I was fourteen years old.

In the dead of winter with the temperature quite low, and the snow continuing to fall, I was in a battle at a place called Valley Forge. I had no coat or shoes that were fit for the type of weather I was facing. To make matters worse, our food supply was continuing to dwindle and the necessary ammunition for our weapons was also becoming lower and lower. With the fierce raging cold beginning to take its toll, I began to freeze. It was not long until I found myself slipping into a deep darkness. My bodily heat had dissipated completely and my entire body was frozen stiff. I gave my life so others may live in freedom. I was fifteen years old.

During the Civil War, I was sent to a place called Gettysburg, where again we encountered much hostility. The opposing army was determined to defeat me but I was equally determined to survive. As our unit crossed a wheat field, I suddenly was hit in the chest by a minie ball. The impact was so great, I fell to the ground and lay there watching my companions move on beyond me. I couldn't move. My life was rapidly passing. I breathed my last breath. I gave my life so others may live in freedom. I was sixteen years old.

In the year of 1918, I was sent to France as a part of an infantry unit. I had received my assignment to assist in the war that was raging through Europe. The German army had entered this land with well-trained men and adequate weapons in their attempt to dominate the entire continent. Against these aggressors, we were deployed to try and withstand their attack. With their airplanes, they were dropping numerous bombs and were destroying many lives. In a bitter battle one night, I received several serious wounds to my head and chest. I lived through the night but as the sun rose the next morning the light within me went out. I gave my life so others may live in freedom. I was fourteen years old.

At a place called Normandy, in the year of 1944, our unit was among the first to make landfall after we had crossed the English Channel. The day was June 6, and the number of allied personnel on the ships, and in the airplanes, numbered into the thousands. This was to be the greatest invasion ever recorded in history. The Nazi soldiers were entrenched in their bunkers all along the coast of France. They were equipped with powerful weapons and an endless amount of other resources to defend this land against our forces. As I entered along with the others, it was our goal to defeat this enemy regardless of the cost. I didn't get too far after leaving the Higgins boat that I had ridden from the ship, until I was hit. I kept moving inland, trying to ignore the pain of my injury, but then, I was hit again. This time, I couldn't go much farther. There were at least five holes in my body from which blood was flowing. I died in these bloody sands. I gave my life so others may live in freedom. I was fifteen years old.

We were deployed by ship to a Pacific island called Iwo Jima which had been taken by the Japanese army. Our orders were to attack and destroy and liberate the island. We landed without incident on the beach, and started inland. As we moved deeper into the area, suddenly we were ambushed by the enemy. They had hidden in the caves to await our entrance. We found ourselves in a crossfire as we were heavily assaulted by an enemy that we could not even see. I ran forward but then, a bullet struck me from the rifle of a well-hidden Japanese sniper. In the sands of Iwo Jima, I lost my life. I gave my life so others may live in freedom. I was sixteen years old.

On a perfectly beautiful evening, we climbed aboard our aircraft and soon were flying through the sky. We had received orders to make a bombing run on a fuel depot in order to destroy the oil tanks as well as the railroad line. I was a gunner and my training had prepared me to seek out the targets below and destroy them. We successfully scored a number of direct hits. As we prepared to leave the area, suddenly our plane was struck with a large shell. The damage was so severe that we could not continue to remain airborne. As we lost altitude, the turret in which I was positioned, was blown apart by enemy fire, my life ended. I gave my life so others could live in freedom. I was sixteen years old.

I served aboard ship in the Atlantic and in the Pacific and during my tenure we encountered a number of engagements with both German and Japanese ships. We were successful from time to time, but on one occasion, an enemy submarine scored a direct hit on our ship. We continued to do battle but then another torpedo found its way to our ship and did extensive damage. The call to abandon ship was sounded and men began to jump into the briny deep. As I ran across the bow of the ship I was hit with a piece of shrapnel. It was a death blow, so I fell to the deck lifelessly. I gave my life so others may live in freedom. I was fifteen years old.

In the early fifties, I was assigned to a military unit in the land of Korea. The North Korean army, backed by the Communist forces of China was attempting to bring the entire land under the dictatorship of an evil regime. Along with my companions we were sent to a place called Pork Chop Hill to engage the enemy. Of all the wars in which I had participated, this one was quite different. There were times that we were not sure as to who the enemy was. The one thing we did know was that the people we were facing were cruel and dedicated to their evil intentions. We fought as long as we could but we were beaten. I was captured and taken to their leaders. The brutality that they inflicted in torturing me was almost beyond belief. But somehow, I survived as a POW for several months. One night, after days of planning, I chose to make my escape from these heartless captors. I almost made it, but then, a bullet brought me down. I gave my life so others may live in freedom. I was fourteen years old.

In the jungles of South Vietnam, we were assigned search and destroy missions. The specific location that we were assigned to penetrate was the Pleiku area. It was here that we were ambushed by a large North Vietnamese unit. During the battle, I was hit by gunfire four times. I fell to the ground and was bleeding profusely. When the medics arrived, they saw that my life had ended. I gave my life so others may live in freedom. I was sixteen years old.

There are hundreds of cemeteries in various parts of the world where literally thousands of soldiers, marines, sailors and air force personnel from the Army, the Navy, the Marines and the Air Force have been interned. Their bodies were never sent home for proper burial due to the horrendous circumstances of their deaths.

It is estimated that in excess of 200,000 men and women had enlisted in the military before they reached the age in which they would have been considered adults. Even though age-wise, they would have been considered juveniles, their dedication and their courage, and especially their sacrifices proved them to be mature in the most meaningful way. Today with our advanced computer technology it is virtually impossible for underage individuals to enter as these did at an earlier time.

Those who are in Iraq and in Afghanistan are there as I was and as all these others were who have gone on. These military personnel are doing what they are doing for the same reason we did, to offer their lives so others, even you and I, may live in freedom.

(Written as a tribute to all underage military personnel
of all the wars that have been fought)


Ophelias "Joe" Bourque

# The Background

*(Exerpt from America's Young Warriors -*
*By Ray D. Jackson and Susan M. Jackson*

*"... I did nothing spectacular nor made any contribution to my country or flag — except love the hell out of them — still do after 50 years."*

This was part of the response of an underage veteran when asked if he would allow his story to be published. His statement exemplifies the attitude of the great majority of the men and women who chose to serve their country before they had attained legal age. Most of them play down their accomplishments, which in some cases are legendary, but all of them express a strong love for the country which they so proudly served.

The quote above was from a veteran of World War II who had first joined the U. S. Army at age 14 and later served in the U. S. Navy. During the World War II era, literally thousands of men and women circumvented the age requirements in order to serve in the military. Most completed their enlistments without ever being discovered to be underage. Of those whose real age was revealed, some were allowed to remain in the military, while others were given honorable discharges and sent home. A few received bad-conduct discharges from officers who were determined to "teach those kids a lesson." Punishment seemed to depend more on who was interpreting the law than on the law itself.

For years, most of these veterans talked little about their underage exploits. Even five decades later, some were still afraid that the government would punish them if their underage enlistments were to be discovered. To this day, there are a few who will not reveal their underage service for fear of retribution.

In 1978, a small group of underage veterans started an association called the Underage Servicemen of America. While it was in its embryonic stage, the founder died and the association folded. It wasn't until 1991, when Allan Stover of Ellicott City, Maryland, began an association called the Veterans of Underage Military Service (VUMS), that men and women who had served underage began to receive recognition. Letters were obtained from all branches of the armed forces stating that no punitive action would be taken against those who served under the legal age limit. With these assurances, the association initiated an aggressive campaign to advise underage veterans that it was safe to "go public." Since its incorporation in 1991, the association has identified over 1000 underage veterans. Over 1302 men and women are active members of the association at the time of this writing (January 2010). This is a small fraction of the estimated 100,000 underage veterans who served during World War II, the Korean War, and the war in Vietnam.[1]

One of the purposes for which the VUMS was formed, as stated in an early edition of the association newsletter, was "... to record for history the experiences of that unique group called underage veterans." This book is an attempt to record some of that history.

Before proceeding to the legacy of underage veterans and to their stories, some background information is needed to better understand the terminology and the sequence of certain events.

An underage veteran is generally considered to be one who served in the armed forces before reaching the age of 17. However, age requirements did not become uniform among the various branches of the military until 1948. Before then, the age requirements depended upon the branch of service and the gender of the applicant. The Navy, Marine Corps, and Coast Guard allowed enlistment at age 17 with parental consent and at age 18 without consent. Apparently, this has been the policy for most of this century.

During the early months of World War II, Army policy required enlistees to be 18 years old, with no option for 17-year-olds to join with parental consent. In fact, parental consent was required until the age of 21 if the choice was one of the more "hazardous" branches of the Army, e.g., the Air Forces or the paratroops. In 1948, all branches of the services adopted the policy of allowing enlistments at age 17 with parental consent and at age 18 without. This policy also applied to women.

Women have served in the armed forces of the United States since the Revolutionary War and in every war since that time. Many were doctors and nurses. Records of women serving underage during the early years of our country were either nonexistent or not readily accessible. During World War II, with the establishment of women's service branches, a number of women circumvented the system and served underage.

The Women's Auxiliary Army Corps was established on 15 May 1942. Women enrolled in the WAAC were technically not considered soldiers until 1 July 1943 when the Women's Army Corps (WAC) was established. The Women's Reserve of the U. S. Navy (WAVES) was established 30 July 1942, and the Marine Corps Women's Reserve was established on 13 February 1943.

Age requirements for women in all branches were age 20 with parental consent and age 21 without consent, until passage of the Women's Armed Services Integration Act on 12 June 1948.

In addition to the WAAC changing to the WAC, another name change has created some confusion. Throughout World War II, one of the most popular military songs included the words, "Nothing will stop the Army

---

[1] Ray D. Jackson, "Rushing the Cadence — Serving Underage," Leatherneck, pp. 52-53, Dec. 1996.

Air Corps." Later in the war, it became, "Nothing will stop the Army Air Force," with punsters adding, "except the weather." Actually, both names, "Corps" and "Force," were wrong. When airplanes were first used by the U. S. Army, they were assigned to the Signal Corps. Later, the U. S. Army Air Corps was formed, and that name was used until 20 June 1941. On that date, the Army Air Corps officially became the U. S. Army Air Forces, with an s. On 18 September 1947, the U. S. Air Force was formed as a separate service.

## The Legacy

Many young men and women who circumvented the system and entered military service underage during the World War II — Korean War era thought that they were the only ones, or one of a very few, who had ever done so. With the founding of the Veterans of Underage Military Service (VUMS) in 1991, most were surprised to find that hundreds of others had also served as youngsters. In fact, the legacy of underage service in the United States military goes back at least as far as the Revolutionary War.

During the American Revolution, Washington's army was comprised of volunteers who simply appeared on the scene and augmented the numbers of regular militia companies in which every man between the ages of 16 and 60 was expected to serve. Records of underage veterans of that war are scarce, but there is little doubt that boys younger than 16 served in Washington's army. Fortunately, the deeds of two youngsters who served as warriors and statesmen, and helped shape the nation, have been documented.

Jonathan Dayton joined the 3rd New Jersey Regiment in 1775 as an ensign at age 15. This was arranged by his father, the regimental commander. At 16, Jonathan was a lieutenant and took part in the heavy skirmishing between Washington's main army and British forces threatening Philadelphia. He also participated in the battles at Brandywine Creek and Germantown and endured the cold and hunger at Valley Forge. By 1780, Jonathan was a captain. He transferred to the 2nd New Jersey Regiment, took his company to Virginia, and fought in the decisive Yorktown campaign.

After the war, Jonathan assumed responsibilities in the family's mercantile business and studied law. He entered politics and represented Elizabethtown in the legislature. When New Jersey selected delegates to the Constitutional Convention in 1787, Jonathan Dayton was appointed a delegate. He also had interests in the Ohio region. Later, a town in Ohio would be named for him.[2]

William Jackson was born in England in 1759. While he was a young boy, his parents died and neighbors arranged for his emigration to

Charleston, South Carolina, to be reared by a family friend. His benefactor commanded the Charleston Battalion of Artillery. William attended many musters of the battalion and obtained an appointment as a cadet. The date of his entry into the 1st Carolina Regiment is not known, but it may well have been prior to his seventeenth birthday.

Jackson fought in a number of skirmishes during the war and was captured when the British took Charleston in 1780. He was exchanged and became a secretary to one of Washington's aides, and in 1782 served as Assistant Secretary at War. In 1787, he applied for the position of Secretary of the Constitutional Convention and edged out Benjamin Franklin's grandson for the post. After the convention, he was selected by George Washington to be his secretary, becoming one of the original civil servants in the executive branch.[3]

The patriotism and sense of adventure of youngsters who wanted to serve their cause during the Civil War is much better documented. By this time, record-keeping was more advanced and letters from the soldiers themselves have been preserved.

The Civil War may well have been called "The Boys' War." Authorities differ on the ages of soldiers in the Union armies, but it has been reported that of the 2,700,000 men in blue, about 800,000 were seventeen or under, about 200,000 were sixteen or under, about 100,000 were fifteen or under, about 300 were thirteen or under, and about 25 were ten or under. Most of the young soldiers slipped in as musicians. There are many tales of buglers who were too small to climb into saddles without help, yet rode into pistol-and-saber battles with their regiments.[4]

Johnny Clem was a drummer boy at age 9.[5] At age 11, he was with the 22nd Michigan Regiment and became a mounted orderly on General George H. Thomas' staff, with the rank of lance sergeant.[3] Later, during peacetime, he became a major general.[4]

The First Minnesota Volunteers was a regiment formed in 1861 in response to President Lincoln's call for volunteers. One of the first applicants was Charley Goddard, a lanky and mischievous youth of 15 from Winona, Minnesota, who told everyone he was 18 and got away with it. Goddard's mother was opposed to his enlistment from the beginning. Apparently, she suggested in her letters to him that he should come home. His response was, "Jest think how I would be received by friends if I deserted and not only that but it would disgrace the name of Goddard." Charlie wrote to his mother often, and fortunately, his letters

---

[2] U. S. Army Center of Military History, Bicentennial Series Pub. 71-19.

[3] U. S. Army Center of Military History, Bicentennial Series Pub. 71-23.

have been preserved.

During the Battle of Bull Run, his regiment was on the line, but Charlie was in a Washington hospital. His illness nearly ended his army career. "I was laying on the bed with my cloths on & there was a doctor come in one that did not belong to my ward," Charlie wrote his mother. "He asked me how old I was I supposed he only wanted to know to satisfy himself so I told him I was 17 years he did not say any thing but continued to make his rounds from room to room and the first thing I knew he handed me a discharge I took it and saw what it was I said nothing but maid for the Regiment and told Capt H C Lester I did not want a discharge he told me I need not take it unless I wanted to this is the way the thing was arranged."

Charlie was wounded in the leg and shoulder on Cemetery Ridge during the Battle of Gettysburg, 2 July 1863. He was discharged at age 18 and returned to Winona to try to support his mother. He was increasingly plagued by his wounds and the illnesses that had followed him through the war. Nonetheless, he decided to seek public office and won the Republican nomination for county registrar of deeds. He also won the election, but a few weeks later he was dead at 23 years of age.[6]

The Confederate Army also had many underage soldiers, including:

George S. Lamkin of Winona, Mississippi, who enlisted when he was 11 and was severely wounded at Shiloh before his twelfth birthday.

E. G. Baxter, of Clark County, Kentucky, who is recorded as enlisting in the 7th Kentucky Cavalry when he was not quite 13, becoming a second lieutenant a year later.

John Bailey Tyler of the 1st Maryland Cavalry who was 12 when the war came. He fought with his regiment throughout the war without being wounded.

T. G. Bean of Pickensville, Alabama, who organized two companies of troops at the University of Alabama in 1861 at age 13. However, he did not serve until two years later, at age 15.

M. W. Jewett of Ivanhoe, Virginia, who is reported to have joined the 59th Virginia Regiment at age 13. He served at Charleston, South Carolina, in Florida, and at the siege of Petersburg.

W. D. Peak of Oliver Springs, Tennessee, who enlisted in the 26th

---

4 Burke Davis, *The Civil War, Strange & Fascinating Facts*, The Fairfax Press, 1982. Originally published as *Our Incredible Civil* War by Holt, Rinehart and Winston, Inc.

5 David Herbert Donald, *Living Through a Civil War, in We Americans*, a series by the National Geographic Society, 1975,1988.

6 Richard Moe, *The Last Full Measure - The Life and Death of the First Minnesota Volunteers*, Avon Books, New York, 1994.

Tennessee Regiment at age 14.

Matthew J. McDonald who served with the 1st Georgia Cavalry at age 14.

John T. Mason of Fairfax County, Virginia, who served with the 17th Virginia Regiment at the first battle of Manassas at age 14. Later, he was a midshipman on the Shenandoah, a Confederate Navy cruiser.

Billings Steele, a grandson of Francis Scott Key, who crossed the Potomac to join Colonel John S. Mosby's Rangers at the age of 16.3

Six boys received the Congressional Medal of Honor for their exploits during the Civil War. These six are listed below, followed by their age at time of deed, not their age at time of enlistment

Willie Johnston — 11 years and 11 months.

John Angling — 14 years and 2 months.

Oscar E. Peck — 14 years and 5 months.

Orion P. Howe — 14 years and 5 1/2 months.

John Cook — 15 years and 1 month.

William H. Horsfell — 15 years and 2 months.[7]

It would be eighty years after the Civil War before another youngster under the age of 18 would be awarded the Medal of Honor. Six days after his seventeenth birthday, Jackyln Lucas, who had joined the U. S. Marine Corps at age 14, covered two Japanese hand grenades with his body, saving the lives of other Marines near him.[8]

During times of war, it is natural for youngsters to want to participate, as evidenced by the large number of young soldiers who participated in the Civil War. But between wars, the military was attractive as a means for young people to be on their own and satisfy their sense of adventure.

In 1897, 15-year-old Jones Morgan ran away from home and enlisted in the U. S. Army. He was assigned to the 9th Cavalry and became a Buffalo Soldier. He watched as Buffalo Soldier units joined Teddy Roosevelt's Rough Riders in their charge up Cuba's San Juan Hill on 1 July 1898. Men of the 9th Cavalry were the first to go over the top of the hill. His parents found him soon after that, and his military career ended. He had served two years. During a 1992 interview at his home in Richmond, Virginia, 109-year-old Jones Morgan recalled his days with the 9th Cavalry, "They used me for domestic stuff. Anything that needed fixing in the place, they sent me".

He died shortly after the monument to Buffalo Soldiers was dedicated in Fort Leavenworth, Kansas, 25 July 1992.[9]

Smedley D. Butler became a second lieutenant in the U. S. Marine Corps at age 16. He was born in West Chester, Pennsylvania, on 30 July

---

[7] Congressional Medal of Honor Society, West 46th Street and 12th Avenue, New York, New York.

[8] Edward F. Murphy, *Heroes of WWII*, Ballantine Books, New York, 1991.

1881, the son of Thomas S. Butler, a 32-year veteran of the U. S. House of Representatives. After the USS Maine was blown up in Havana harbor in February 1898, Smedley tried to join the 6th Pennsylvania Volunteers, but was politely told to "run along home." He tried to join the Navy, but his father refused to give his consent. His father, who was then chairman of the House Naval Affairs Committee, returned to Philadelphia one evening and Smedley overheard him tell his wife, "Today Congress increased the Marine Corps by twenty-four second lieutenants and two thousand men for the period of the war." Later, Smedley told his mother that he was going to be a Marine, and if she didn't agree, he would run away and join a regiment where he wasn't known.

The next day, Smedley and his mother went to Marine Corps Headquarters in Washington. They spoke directly to Colonel Commandant Heywood. The Commandant said, "When I met your father the other day, he told me you were only sixteen." "No sir," Smedley lied, "That's my brother." The commandant asked, "How old are you then?" Smedley replied, "I'm eighteen, sir." He was accepted in the Marine Corps as a second lieutenant and retired in 1931 as a lieutenant general. He died in 1940.

Smedley Butler's thirty-three years in the Marine Corps is the source of many legends. The Marines called him "Old Gimlet Eye." He is one of very few men who have been awarded the Congressional Medal of Honor twice. The destroyer USS Butler, later a high-speed minesweeper, was named for him.[10]

Nathan E. Cook talked his sister into signing papers that allowed him to enlist in the U. S. Navy in 1901. Nathan, born in Hersey, Michigan, on 10 October 1885, was 15 years old. He saw action in the Boxer Rebellion in China, the Pancho Villa Mexican-border skirmish in 1917, World War I, and World War II. During World War I, Nathan was in command of a subchaser. His ship sank two German submarines off the coast of Greece.

Nathan's most memorable experience occurred while serving aboard the USS Kansas in 1907. As the ship was leaving the port of Gibraltar, his appendix burst just as he picked up a hawser (heavy mooring line). The Navy doctor thought he was dead and notified his wife that they were sending his body home for burial. He was packed in ice, but he revived en route. Surgery was performed in Philadelphia, and he completely recovered.

Nathan Cook served forty-four years in the Navy. He died at the Veterans Medical Center in Phoenix, Arizona, on 10 September 1992, at the age of 106 years and 11 months. He is buried in the Arizona Veterans

---

[9] Adapted from an article in the *Columbus Ledger-Enquirer*, Columbus, Georgia, 21 July 1992.

[10] Lowell Thomas, *Old Gimlet Eye*, Fairer and Rinehart, New York, 1933.

Cemetery, Cave Creek, Arizona.[11]

World War I again fired the minds of young people, and many manipulated the system to enlist.

Mike Mansfield left his Great Falls, Montana, home in 1917, and on 23 February 1918, at the age of 14, enlisted in the U. S. Navy for the duration of World War I. In 1919, while still 15 years old, he joined the U. S. Army for one year. In November 1920, he joined the U. S. Marine Corps and served for two years.

After his honorable discharge from the Marines, Mike returned to Montana and worked as a mucker and mining engineer until 1930. Although he had never completed grade school or high school, he was admitted to the Montana School of Mines in Butte, Montana, after completing a qualifying examination. He met and later married a Butte high school teacher, Maureen Hayes. With Maureen's academic, financial, and moral support, Mike transferred to the University of Montana at Missoula to continue his education. Later, he taught Latin American and Far Eastern history at Montana State University.

In 1942, Mike Mansfield was elected to the United States Congress and served five terms as Representative of Montana's 1st District In 1952, he was elected to the United States Senate. In 1961, he was elected Majority Leader of the Senate and served in that position until his retirement in 1977 — longer than any majority leader in the history of the United States Senate.

Senator Mansfield served five presidents in a number of capacities. His final assignment was Ambassador to Japan, an office he held from 1977 until 1989. He and his wife currently reside in Washington, D.C.

The Depression years between World War I and World War II provided incentive to a number of youngsters to enlist in the military. The onset of World War II ignited patriotic fervor throughout the nation and was an impetus for underage men and women to join the military in numbers not seen since the Civil War. One of these was a young man from Callicoon, New York.

Alien Clifton Heyn falsified his age and joined the U. S. Navy soon after the attack on Pearl Harbor. He was just 16 years old. He was soon one of 700 men aboard the USS Juneau heading for the South Pacific. Alien was the youngest crew member on the ship, and in the eyes of his shipmates, the least likely to survive a disaster at sea.

Two of Alien's friends were George and Al Sullivan. Along with their brothers Frank, Matt, and Red, the five Sullivan brothers had received special permission from the Navy to serve together aboard the Juneau. On the second day of the Battle of Guadalcanal, at 11:01 a.m, Friday, 13

November 1942, the Juneau, already suffering battle damage, was struck by a torpedo launched from a Japanese submarine, the 1-26. A massive explosion rocked the ship and she quickly sank. Of the crew, 560 were killed in the explosion and 140 survived. Thinking that no one could survive such an explosion, and concerned that the submarine that launched the torpedo would soon strike again, the remaining ships were ordered to steam away and not search for survivors.

The survivors, clinging to rafts, had no food or fresh water. Sharks soon were circling and started taking their toll, attacking the men who were clinging to the rafts. Each day, fewer men were left. Of the 140 who survived the sinking, only ten are known to have reached safety. Alien Heyn, one of the ten, was with one of the largest groups of survivors, which included George and Al Sullivan. Alien, the sole survivor of that group, was alone on a raft when he was finally rescued, eight days after the explosion that blew him off the ship.[13]

The preceding accounts are a very small sample of the legacy of underage veterans. Undoubtedly, the exploits and accomplishments of many more young warriors were just as compelling as these, but most were not recorded and have been lost to history. The remainder of this book is devoted to recording the histories of men and women who served before they could do so legally during the period from World War I through World War II, the Korean War, and the war in Vietnam. For the most part, the stories are written as told by the veterans themselves. Because of today's technology and improved record-keeping, which make it nearly impossible to enlist in the military underage, this group may represent the last of the legacy.

---

[11] Adapted from *The Stars and Stripes - The National Tribune*, 28 Sep. - 4 Oct. 1992.

[12] Biographical sketch furnished by the office of Senator Mike Mansfield, 11 January 1996.

[13] Dan Kurzman, *Left to Die*, Pocket Books, New York, 1994; Edward E. Leslie, *Desperate Journeys, Abandoned Souls*, Houghton Mifflin Company, Boston, 1988.

# THE MEDAL OF HONOR

*Jack H. Lucas, LCM002*
*Age 14 - U.S. Marine Corp*
*Jack's story from, America's Youngest Warriors - Volume II*

Two members of the Veterans of Underage Military Service earned the Medal of Honor during World War II. The late James M. Logan, LM665, joined the Texas National Guard at age 15. He landed at Salerno, Italy, with the 36th Infantry Division in 1943, where he earned the Medal of Honor.

President Harry S. Truman awarding the Medal of Honor to 17-year-old Private First Class Jack Lucas, USMC, 5 October 1945.

Jack H. Lucas, LCM002, (1928-2008) joined the Marine Corps at age 14 and earned the Medal of Honor on Iwo Jima five days after his 17th birthday. Jack's story is in America's Youngest Warriors Volume II, pages 139-143. It is reprinted here after the citation he received along with his medal. Jack was the youngest Medal of Honor recipient in Marine Corps history and the youngest in any service since the Civil War.

The President of the United States in the name of the Congress takes pleasure in presenting the Medal of Honor to Jacklyn Harold Lucas, Private First Class, United States Marine Corps for service as set forth in the following citation:

"For conspicuous gallantry and intrepidity at the risk of his life above and beyond the call of duty while serving with the First Battalion, Twenty-sixth Marines, Fifth Marine Division, during action against enemy Japanese forces on Iwo lima, Volcano Islands, 20 February 1945. While creeping through a treacherous, twisting ravine which ran in close proximity to a fluid and uncertain front line on D-plus-l day, Private First Class Lucas and three other men were suddenly ambushed by a hostile patrol which savagely attacked with rifle-fire and grenades. Quick to act when the lives of the small group were endangered by two grenades which landed directly in front ofthem, Private first Class Lucas unhesitatingly hurled himself over his comrades upon one grenade and pulled the other under him, absorbing the whole blasting forces of the explosions in his own body in order to shield his companions from the concussion and murderous flying fragments. By his inspiring action and valiant spirit of self-sacrifice, he not only protected his comrades from certain injury or possible death but also enabled them to rout the Japanese patrol and continue the advance. His exceptionally courageous initiative and loyalty reflect the highest credit upon Private First Class Lucas and the United States Naval Service."

# Jacklyn H. Lucas, LCM002

*Age 14 - United States Marine Corps*

My father, a Plymouth, North Carolina, tobacco farmer, died when I was 10 years old. I resented his death and I guess I resented a lot of things because I became a real mean kid. By the time I was II years old I was so rambunctious that my mother sent me to the Edwards Military Academy in Salemburg, North Carolina. The discipline at Edwards helped to straighten me out and I became a cadet captain by the time I was 13.

When the Japanese bombed Pearl Harbor, that did it for me. I was determined to join the Marine Corps and fight the enemy. I tried to enlist, but I couldn't fool the recruiters. I turned 14 on 14 February 1942 and was still determined to get into the fight. During the summer I told my mother that I was going to enlist and that I was going to sign her name to the enlistment papers. I told her that she could try to stop me, but if she did, I'd find another way. She agreed not to try to stop me on the condition that I finish my schooling when I was discharged. I went to the Marine recruiting office in Norfolk, Virginia, and told the recruiter that I was 17. I passed all the exams and when they gave me the papers for my mother to sign, I signed them. I was 5-foot 8-inches and weighed 185 pounds, but less than six months past my fourteenth birthday. I was sworn into the Corps on 6 August 1942.

Boot camp at Parris Island was a snap. Three years in a military school had prepared me very well for the Marines. The weather was hot and humid at Parris Island in August and September, so the DI would have me drill the platoon while he stood in the shade of a tree. You can imagine what the troops would have thought had they known that it was a 14-year-old kid drilling them.

After boot camp I went to Camp Lejeune, North Carolina. I did well in the heavy machine-gun school and was detailed to the training command. This was not what I had in mind when I joined the Corps; I wanted to fight. When the rest of my unit was ordered to San Diego I was supposed to stay at Camp Lejeune, but I packed my seabag and got on the back of the train and stowed away to California. A sergeant discovered that I didn't have any records, but it was more trouble to send me back to North Carolina than to keep me, so they kept me. Shortly after arriving in San Diego we sailed for a staging area in Hawaii.

At Camp Catlin on Oahu, I made a mistake. In a letter to my 15-year-old girlfriend in Swanquarter, North Carolina, I mentioned that I was also 15. Apparently a mail censor noticed this and I was ordered to report to the colonel to explain things. Confronted with my letter, I had no alternative but to admit my age. I told the colonel that if I was discharged from the Marine Corps, I would join the Army. I guess the colonel believed me because he didn't have me discharged, but he wouldn't let me go into combat. My unit left for Tarawa, but I was stuck at Camp Catlin.

This was a frustrating time for me. I had enlisted to fight, and here I was stuck in Hawaii. I noticed that men who got into trouble and spent some time in the brig were frequently transferred to a combat outfit. This looked like a way out for me. My dad had taught me to box when I was a kid and I have always been a scrapper. It was time to put my 18-inch biceps to use. In 17 trips to town, I got into 17 fights. I provoked a fight anytime I could and was locked up a number of times. I mashed up a sergeant and was sentenced to 30 days on bread and water.

After I got out of the brig one time, a buddy and I decided to celebrate. We shanghaied a truckload of beer from ships stores and treated our company. The troops had worked their way through ten or twelve GI cans full of iced brew and it looked like we needed some more, so a pal and I went back for more. We were too inebriated to operate efficiently and the police caught us.

While I was in the brig after the beer-hijacking incident, I decided that this course of action wasn't getting me anywhere, except the brig. While listening to the radio I heard Tokyo Rose say that troopships at Pearl Harbor would soon be heading for Iwo Jima. So on 9 January 1945, I packed my seabag, went to Pearl Harbor, and caught a Higgins boat that was headed for the USS Deuel CAP A- 160). I was lucky; my cousin Samuel Oliver Lucas was aboard. With Sam's help, I hid in landing craft and slept on the weather deck for 29 days. I knew that I would be listed as AWOL for 29 days, but after 30 days, I would be considered a deserter. That day I turned myself in to Captain Robert H. Dunlap, Sam's company commander. Captain Dunlap took me to Colonel Pollock and told him the situation. The colonel said to me, "I'd like to have a whole shipload of Marines that want to fight as bad as you." I was assigned to C Company, 1st Battalion, 26th Marines, 5th Marine Division, Captain

Dunlap's company. I did not inform the colonel that I was 16 years old.

The ship stopped at Saipan and a Marine was taken ashore with appendicitis. I was issued his rifle and gear. On 14 February 1945, I turned 17, but there was no party for me on the Deuel. On 19 February we hit the beach at Iwo Jima. Shells were flying, people were being blown apart, and bullets were everywhere. This was just where I wanted to be. I was as anxious as ever to fight and kill as many of the enemy as I could. During the first two days, we suffered 5,370 casualties, killed and wounded.

The next day we were making our way toward the Japanese airstrip on the plain northeast of Mount Suribachi. We stopped to pound an enemy pillbox and jumped for cover into one of two parallel trenches that led to it. To our surprise there were eleven Japanese in the other

*Jack Lucas stood in the shade of a tree. You can imagine what the troops would have thought had they known that it was a 14-year-old kid drilling them.*

trench. They were too close to put our rifles to our shoulders so we opened fire offhand. Just after I shot the second enemy soldier my rifle jammed. I looked down to try to get the damned thing to work when I saw the grenades. I hollered, "Grenades!" and dove for them. I smashed my rifle butt against one, drove it into the volcanic ash, fell on it, then pulled the other one under me. I was there to fight, and we were there to win. You do what you have to do to win. It was not in me to turn and run. The volcanic ash and the good Lord saved me.

If I'd been on hard ground that thing would have split me in two. There was just one explosion, but one was all I could handle, and I had trouble handling that one. It blew me over on my back, and it punctured my right lung, but it never knocked me out. It also got me in the thigh, neck, chin, head, chest, right arm, and hand.

The rest of the team sprinted down the trench, turned and fired down the other trench, killing the Japanese soldiers. I was lying on my back with my right arm twisted so far underneath me that I thought it had been blown off. My mouth and throat filled with blood and I probably would have drowned if I had lost consciousness. My left hand was the only thing I could move and I kept moving it to show that I was alive. A Marine from another unit came up and I was afraid I would be shot because I was barely recognizable as an American. The Marine called for a corpsman and while he was working on me, a Japanese soldier popped up from a hole in the trench. The corpsman shot him and continued to work on me.

A mortar barrage on the edge of the trench delayed the stretcher bearers. When they finally carried me off, one stumbled and dropped his end of the stretcher. I split my head open on a rock. I looked up at him and smiled to let him know that I knew what he was trying to do, and that I appreciated it. I could see that he was exhausted.

On the evacuation beach a corpsman covered me with a poncho for shelter from the elements. I thought, "Oh, Lord, I'm dead!" Then the morphine took hold and I passed out. As they were hoisting me aboard an LST they nearly dropped me into the sea. Someone caught my foot and pulled me aboard. I was put in the hold of the LST with hundreds of other wounded until there was room for me on the hospital ship USS Samaritan. Before the ship sailed for Guam the American flag went up on Suribachi. The word was passed throughout the ship and I felt as jubilant as I could be. I had fought for my country and I felt a great deal of pride in that. My only regret was that I didn't get to stay there longer to kill more Japs.

Aboard the hospital ship a surgeon said, "Maybe he was too damned young, and too damned tough to die." I went under the knife twenty-two times, and there are still about 200 pieces of shrapnel in my body, some the size of 0.22 bullets. The only time I was ever knocked out in my life was when they gave me an anesthetic. After seven months in hospitals, I was in good enough shape to be discharged.

Jack Lucas received the Medal of Honor from President Harry S. Truman on 5 October 1945. When the President said to him, "I'd rather be a Medal of Honor winner than President of the United States, " Jack replied, "Sir, I'll swap with you." Jack earned the Medal of Honor at the age of 17 years and 5 days. He was the youngest U S. Marine ever to be awarded the nation's highest honor, and the youngest recipient since the Civil War.

Captain Dunlap, who took Jack into his company, also earned the Medal of Honor on the same day that Jack did.

Jack also received the Purple Heart for his wounds. He still sets off metal detectors in airports from the shrapnel he carries in him. True to his word to his mother, Jack finished high school and graduated from college.

During the Vietnam War, Jack accepted a commission in the Army and served for six years, leaving the Army as a captain in 1965. In

1985 Jack returned to Iwo Jima. In his words, "I wanted to walk on the beach without having to duck."

In 1995, he served two tours as a captain in Vietnam. During the State of the Union address, President Clinton recounted Jack's heroic action on Iwo Jima and introduced him, along with a son and grandson to the joint session of Congress which gave him a standing ovation. He has four sons, a daughter and seven grandchildren. One son, a West Point graduate.

Jack Lucas lived a long and interesting life despite the pain caused by the grenade fragments he carried in his body. In mid-April 2008, Jack was admitted to the Forrest General Hospital in Hattisburg, Mississippi. The diagnosis was cancer. Jack approached this situation with the same determination he demonstrated on Iwo Jima. He began treatment for the cancer and went home. A short time later he was back in the hospital. Jack died on 5 June 2008 at the age of 80 years. His wife Ruby and other family members were at his side.

Jack's funeral was held on 9 June 2008 at the Hattisburg Lake Terrace Convention Center, Hattisburg, Mississippi. Jim Jackson, M2424, was invited to the funeral. Jim reported that a Marine 3-star general spoke at the funeral. The general said that he was always using Jack's name when recruiting future Marines, as Jack joined when he was 14. He said that Jack wrote him a letter saying that "he had heard the general was using his name a lot and was sending him a package outlining his life and experiences so he would get it right." Jack's book had not been written at that time.

An Air Force 2-star general and Medal of Honor recipient spoke as a representative of the Medal of Honor Society. Three other generals and several colonels were present. Several family members spoke of Jack's devotion to his family. Several spoke of his dedication to the Marine Corps and to his "aggressive spirit." Graveside services were held at Highland Cemetery in Hattisburg.

In 2006, Jack, (with D. K. Drum) wrote the book Indestructible. In addition to more detail about his Marine Corps service and the invasion of Iwo Jima, the book tells of his boyhood and his life after the war. The dedication in the book reads:

For the men and women who served our great nation, especially my brothers in arms who sacrificed their lives in the preservation of liberty, and to my precious wife, Ruby.

# Medal of Honor Citation
# James Marion Logan

James Marion Logan (December 19, 1920 – October 9, 1999) was a United States Army soldier and a recipient of the United States military's highest decoration—the Medal of Honor—for his actions in World War II.

Logan joined the Army from Luling, Texas, and by September 9, 1943 was serving as a Sergeant in the 36th Infantry Division. On that day, he participated in the Allied landings near Salerno, Italy. Logan single-handedly captured a German machine gun emplacement and later killed an enemy sniper. For these actions, he was awarded the Medal of Honor ten months later, on July 5, 1944.

Logan reached the rank of technical sergeant before leaving the Army. He died at age 78 and was buried at the Texas State Cemetery in Austin, Texas.

Sergeant Logan's official Medal of Honor citation reads:

For conspicuous gallantry and intrepidity at risk of life above and beyond the call of duty in action involving actual conflict on 9 September 1943 in the vicinity of Salerno, Italy. As a rifleman of an infantry company, Sgt. Logan landed with the first wave of the assault echelon on the beaches of the Gulf of Salerno, and after his company had advanced 800 yards inland and taken positions along the forward bank of an irrigation canal, the enemy began a serious counterattack from positions along a rock wall which ran parallel with the canal about 200 yards further inland. Voluntarily exposing himself to the fire of a machinegun located along the rock wall, which sprayed the ground so close to him that he was splattered with dirt and rock splinters from the impact of the bullets, Sgt. Logan killed the first 3 Germans as they came through a gap in the wall. He then attacked the machinegun. As he dashed across the 200 yards of exposed terrain a withering stream of fire followed his advance. Reaching the wall, he crawled along the base, within easy reach of the enemy crouched along the opposite side, until he reached the gun. Jumping up, he shot the 2 gunners down, hurdled the wall, and seized the gun. Swinging it around, he immediately opened fire on the enemy with the remaining ammunition, raking their flight and inflicting further casualties on them as they fled. After smashing the machinegun over the rocks, Sgt. Logan captured an enemy officer and private who were attempting to sneak away. Later in the morning, Sgt. Logan went after a sniper hidden in a house about 150 yards from the company. Again the intrepid Sgt. ran a gauntlet of fire to reach his objective. Shooting the lock off the door, Sgt. Logan kicked it in and shot the sniper who had just reached the bottom of the stairs. The conspicuous gallantry and intrepidity which characterized Sgt. Logan's exploits proved a constant inspiration to all the men of his company, and aided materially in insuring the success of the beachhead at Salerno.

James Marion Logan, Medal of Honor recipient, was born in Luling, Texas on December 19, 1920. As a young man growing up during the Great Depression, Logan took many odd jobs in order to help support his family. He swept floors and dove for golf balls in the San Marcos River, until he was able to pass for seventeen. Like many of his generation, Logan would be called to the battlefields of Europe.

On August 4, 1936, Logan joined the Texas National Guard because it would pay him a dollar a week; much more than retrieving stray golf balls.

At the age of fifteen, Logan enlisted in Company L, Luling Guard, and remained with the unit until it was mustered into federal service in 1940.

He was serving as a rifleman in Company I, 141st Infantry Regiment, 36th Infantry Division (Texas National Guard) when his company landed on the beach near Salerno, Italy on September 9, 1943.

In combat, his old friends said, Logan displayed the same toughness of spirit that he acquired during childhood. Professor David Stroud of Kilgore College stated that, “What Sergeant Logan did that day was more than enough for even a million lifetimes . . .That is why we are here.” In a letter to Sergeant Logan, Governor Bush said, “You have earned the gratitude of the nation and of the state of Texas. I honor your courage, and I thank you for your service.”

However, unlike all but a handful of his fellow soldiers, this young man from Luling, Texas would return home with the nations highest award for valor —the Medal of Honor. In addition to this, Logan also earned the Distinguished Service Cross and the Bronze Star for bravery and the Purple Heart for being wounded. However, there was but one more way to recognize his service to not only the nation but to the State of Texas. More than three decades after its official authorization by the 58th Texas Legislature, Technical Sergeant James Marion Logan received the Texas Legislative Medal of Honor on May 30, 1997. Logan was presented the medal at a ceremony in the Texas House of Representatives Chamber.

Sgt. Logan passed away on October 9, 1999, in Kilgore, Texas and was buried in the Texas State Cemetery five days later.

Reprinted from Logan Online, December 17, 2009

# Douglas Allds

*Age 14 - U.S. Army*

My military experience started quite by accident.About the 15th of Dec. 1945 a friend and I was working at Hotel Racine in the bowling ally.

On our break we walked past the Army recruiting Office. One of the recruiters called us in and asked if we wanted to join the army, of course I did. I knew that I had to be at least 17. So thats what I said. Because of my age he told me that my parents must sign. He gave me the enlistment papers to take home and fill out. It was that simple.

Except that my mother was not that easy to convince that it was the right thing to do. Mother finally came around to my way of thinking. The army would be a good place to grow up. She signed.

The next thing was to pass a medical. This took place at Milwaukee, WI. I don't know how a 14 yr. old could pass for 17, but I guess that was not their problem, it went fine, no trouble, but took most of the day. we were sworn in the army on the spot, this was 21/12/1945. After dinner we went by train to Ft. Sheridan, Il. It was late at night by this time, we were issued just the basics, some clothing & bedding.

The first morning in the army was a real wake-up, the barrack NCOIC was a returning S/Sgt. from WW2 either for discharge/reassignment don't know. But he was not worried about anything or anybody. I heard words that I've never heard before.

It took about a week to process into the army, all the IQ test, shots, what ever else had to be done, it took about two weeks before I was assigned to Sheppard Field, TX. for basic training. Before entering the army I worked out with golden gloves boxing team in Racine,WI. I'm not old enough to fight, but the instructor let me train with them, so I'm in good condition. Basic training was a breeze as far as the work went.

After basic, I was assigned to Scott Field, IL. for Radio Operator school, I was there 08/04/1946 Air Force Day. The first time I seen a F-80 shooting Star.

Our Sqd. stood on the ramp seemed like about 2 1/2 Hrs. for the ceremony, lots of people making speech's, I don't think I ever heard a word, in southern Il. it gets very hot for a WI. boy.

After Scott Field it was time for over seas duty. At the time I enlisted one could pick place of duty, I choose Europe. The army sent me to Camp Kernes, Utah. After I got there I realized that was for the Pacific zone. I then told people at the orderly room I was to go to Europe. Back on the troop train for Camp Kilmore, N.J. It was late summer when I boarded the Ship for Europe. Late summer is not the time to be on the Atlantic ocean. I never was that sick before or since. It took 10 days to cross the ocean. The first 4 days I wished that somebody would push me overboard. I did not have the courage to jump. But I did survive.

Germany was a whole new experience. This was right after WW II. The country was trying to reorganize the people had nothing. Much of the country was still in shambles.

We landed in Bremerhaven, went on a train to Firstenfelbrok, this was Gorings Headquarters at one time, very impressive place, barrack's Buildings were all 4 man rooms. Could of been two. Swimming pool was 16' deep. Everything got your attention. I was there about 30 days.

Then was assigned to Augsburg in southern Germany. This was the German experimental AB for Jet aircraft. There was still one that crashed on the side of runway, while there I went To Garmisch for 5 days & took a cable car to the top of the Zugsitze Mt. said to be highest point in Germany.

It was at this time that I got caught up with. It seems that the truant officer came looking to see way I was not at school. Mother told him I was in Germany, and in the army. The C9 Commanding officer called me in his office, first thing he asked was my age, I gave military age, he then asked for the truth. I knew that I was had. He had all the paper in front of him. After that it was just a matter of when.

Men were waiting for their turn to rotate back to U.S.A. The next day I was on my way.

After a Honorable Discharge and all mustering out pay, I was entitled to G.I. Bill and all that goes with it. When I turned 19 during Korea war, 22/12/1950 I reinlisted for 4 more yrs. All in all if I was to do it over I would. Meet great people, that served their country. And am proud to do so.

# Easton J. Ardoin

*Age 14 - U.S. Army*

## Part I

## Ardoin started long career at early age

*Written by Ed Howard*

As with the familiar dream of running away with the circus, many a young boy has dreamed of enlisting in the Army.

For a youngster from the bayous of Louisiana, the dream was strong enough to act on while still a boy. Twice.

At 14, Easton J. Ardoin lied about his age in order to join the Army. His first "enlistment" lasted just over a week, until his secret was found out and he was discharged.

The young resident of Chataignier, La., thought it wise to wait before trying to enlist again. So he did. For a year.

**Finally joins Army**

In 1948, at age 15, Ardoin had a friend forge his parents' signature and once again joined the Army. This time, he managed to get past the reception area and into basic training. There at Fort Jackson, S.C., he succeeded in bluffing his way through the 18-week period. As recounted last week by Fort Sill's commanding general during a special ceremony, the suspicious drill sergeants there attempted to break the teenager's bluff by keeping him on constant KP during the day and having him shovel coal by night. Whether he actually fooled the cadre or not is open to debate. But after he reaffirmed to his commander that he wanted to stay in the Army, the question of his age was dropped.

After advanced individual training - during his first stay at Fort Sill. Ardoin got his corporal's stripes and an assignment to West Point to help train cadets in artillery. Next came a posting to Camp Hokkaido, Japan, in 1949. In addition to military training - including live-fire exercises on Mount Fuji - the school dropout completed his education with high school classes at night.

In June 1950, came the outbreak of the Korean War. Ardoin's unit deployed to South Korea and went into combat on Independence Day. Ten days later, the unit was overrun by North Korean soldiers. Ardoin had to crawl over the corpse of his battery commander to get inside the rim of a knocked-out truck, where he hid from the communist troops. Walking for two days, he joined other scattered troops at Taejon Airstrip, where they formed and ad hoc detachment with a pair of borrowed howitzers.

### Returns to Sill

Ardoin's post-Korean War service included a return to Fort Sill and some other overseas duty tours, including one in Vietnam. He retired from the Army in 1980. Then began his second career at Fort Sill, as a game warden with the post Fish and Wildlife Branch and eventually the job of director of Sportsmen Activities. There, he still works with soldiers, as well as civilians, and derives tremendous satisfaction from the work he does. Much of that comes from being able "to help a young soldier get out of the barracks and get some recreation" from the hunting, fishing and other outdoor activities on post. Another portion, no less satisfying, comes from taking the extra steps needed to help disabled and elderly military retirees enjoy those same activities.

"I'd rather see 'em out on the range than in a nursing home," he said, in the still-heavy Cajun accent which years ago won for him the nickname, "Frenchy."

Asked what possessed him to join the Army at such a tender age, Ardoin cited the inspiration of his Uncle Morris, who served as one of America's first paratroopers in World War II. Initially, he planned to follow his uncle's footsteps - right out of a plane.

"I was going to go to jump school." Ardoin explained. "But I didn"t weigh enough." And his plans to "bulk up" to qualify for paratrooper training were interrupted by that little fracas in Korea.

# Part II

## Korean War vet recalls lessons learned

*Written by Robyn Baer*

He doesn't want to talk about the politics.

Memories of the Korean War are what retired Command Sgt. Maj. E.J. Ardoin wants to pass on to upcoming leadership of the Army.

"We were told on June 25th, 1950, that the war had broken out, and were put on alert," said Ardoin.

"They told us the enemy had bad equipment - single shot rifles. We would be back within 30 days, they said. But the enemy was underestimated by the leadership - from top to bottom.

With the rest of the soldiers in 63rd Field Artillery Battalion, 24th Division Artillery, Ardoin was shipped from Camp Hakata, Japan, to Chonan, South Korea. Riding small boats with guns he called LSTs, A Battery arrived in Chonan July 4, 1950.

"We took positions, but were run out the first day. We retreated south. Sometime about July 14, I think, we were attacked. It was about 12:30 in the afternoon when we were run over. The enemy had motor and ground troops. There were thousands of them running into the battery area. We were outnumbered 10 to 1. All of our equipment was destroyed. Our battery commander was killed. We had just received replacements the day before, and I never even knew their names."

Ardoin said the soldiers of A Battery, 63rd FA each had only one personal weapon and a couple of "banana clips" of ammunition, and their ammunition trucks were all on fire. The U.S. soldiers destroyed the firing licks in their howitzers and threw grenades in the tubes so the enemy couldn't use their weapons against them. With insufficient ammunition to fight the enemy, said Ardoin, most of his unit was killed or captured that day in Korea.

"Me and a couple of guys from my unit went over a cliff to hide from the enemy," Ardoin said.

"We hid during the day and moved during the night toward the friendly line. After about three days, we regrouped at Taejon Airstrip, where we were overrun again."

He said the enemy had T34 tanks with troops. The soldiers in his unit had no anti-tank weapons, just 2.36 Bazookas, he said.

Ardoin took cover under a bridge while the enemy went over.

"After that, everyone was running - retreating to the South. I didn't belong to anyone. I didn't have a unit. There were only a dozen or so of us left."

Ardoin said in August the soldiers regrouped in Mar Yang.

"We were all sick - malnourished and had diarrhea. We became a part of C Battery, 13th FA, 24th Division. For the rest of the war, we moved north. It was November, and we had just been told we would be home by Thanksgiving. Just when North Korea was about whipped, the Chinese came in."

Ardoin said the winter of 1950 in North Korea was a gruesome one.

"It was 30 to 40 degrees below zero. All we had to sleep with was the old blanket-type sleeping bags. Our combat boots were old and worn out. Then we got snow pack boots, but they didn't help. They would freeze up on you. We had no water-proof clothing whatsoever.

"There was a lot of frostbite, and if you were wounded, you probably froze to death because there was no way to get you out. All we could do was make lean-tos with gun tarps and sleep really close," Ardoin said.

Somehow he got through it.

During the Korean War, Ardoin moved up in the ranks quickly, from a corporal to a master sergeant in just eight months. "Most of our leaders had been killed. I extended my tour three months to become a section chief and then a platoon sergeant," Ardoin said.

After the Korean war, he spent 29 years and six months serving as a first sergeant and sergeant major before retiring here after 33 and a half years of service.

A supervisor and game warden with Sportsman Services, Directorate of Environmental Quality, Ardoin received an award for 50 years in federal service in 1998.

"We learned many lessons in Korea," Ardoin said. "The Army was too small. We didn't have the right equipment. Our Reserve replacements were not fit and not properly trained. We were outnumbered and outgunned the whole way." Ardoin said he has confidence that if the United Stated entered another war like Korea, it would come out on top.

"All we have to do is get back to the facts and the lessons learned from the wars we have fought. We need more cannon artillery. The Army has the best equipment in the world, but we need more conventional weapons to fight a war like Korea again.

"The American soldier, with the proper leadership and training, is a tough soldier who can do anything," Ardoin said.

# Paul Ash, LM698

*Age 16 - United States Army*

I will never forget 7 December 1941. President Roosevelt announced on the radio that Japan had bombed Pearl Harbor. My age was 11 years and 1 month. However, I knew then that I wanted to be in uniform. I wasn't particular, Army, Navy, or the Marines.

As the war progressed, my desire to serve grew. In 1946, just past my 16th birthday, I talked to two of my friends who were both 16 years old into joining the Army with me. I prepared birth certificates and parental consent forms so that we could join the regular Army. In January 1947, we started on our journey to enlist. We made the applications at 39 Whitehall Street, New York City. As we were being processed, a top sergeant put a very large book on his desk that was titled Births of New York City. I thought that the sergeant would know that we were underage and send us away. I told my friends that it was not a good time to consummate our enlistment. We took all the papers and left.

About two weeks later, I asked my friends if they were willing to go back and try again.

When they decided not to go, I decided to go it alone. My parents were moving to Tucson, Arizona, and I thought it would be helpful to them if I went off on my own and not be a liability after their move to Tucson. I was midway in my junior year at Samuel Gompers High School in New York City.

Shortly after I revisited the recruiting station, I took a physical. The doctor told me I was color blind and had flat feet, but that I was still acceptable. I was sworn in and soon I was off to Fort Dix, New Jersey, for infantry basics. It was extremely cold in Trenton, New Jersey, particularly lying on the ground doing dry runs on the rifle range. The Army gave me all the routine intelligence tests and, fortunately, told me that I was eligible for a service school. I told the officer that I wanted to go to the CIC (Counter Intelligence). I thought the officer was going to laugh hysterically, but instead he said very seriously that I would be eligible for the CID (Criminal Investigations Detachment) school at Carlisle Barracks, Pennsylvania. After basics, I was sent to Carlisle Barracks and I completed the U. S. Army CID training.

I requested duty with the First Cavalry Division in Japan, but instead I was transferred to the Brooklyn Army Base not far from my home in

Manhattan. There I served as an enlisted military police on the Brooklyn waterfront docks near the base. I was also given honor-guard duty during military funerals and fired the last salute. I was asked to conduct the monthly troop education program. My first speech was "What is American Democracy." For a kid from the lower East Side of New York, I was given a lot of responsibility, and I loved it.

My next assignment was to be a body escort for soldiers killed in Europe. I did two such missions. During the first mission, my job was to escort approximately 40 bodies to a post near their respective homes. My second mission was to escort one soldier's body recently killed in an automobile accident in Europe. I escorted his body to his home in the Big Thicket near Houston, Texas. The family asked me to appear at the funeral. I obliged and stood at parade rest at the foot of the coffin. The minister conducted the service at the grave side.

When he finished, I folded the flag that was draped over the coffin, with the assistance of an American Legion member. I approached the mother of the soldier and started to present the flag by stating, "On behalf of the President of the United States, Commander in Chief..," and then I choked up, partly because of the heat but mostly because of the solemnity of the occasion. The family was very kind to me. I was still only 16 years old. After the services, the family drove me to Fort Worth and dropped me off at Carswell Air Force Base where I hitched a ride on a B-25 to Davis-Monthan Air Force Base in Tucson, Arizona, to visit my mother and father.

As a 16-year-old boy, the U. S. Army gave me responsibilities that I never imagined I would have. Subsequently, my next assignment was as a stockade guard at Fort Kilmer, New Jersey. I did not enjoy that job, but soon thereafter my enlistment term was up. I decided to get discharged. While in the Army, I obtained my high-school equivalence diploma and earned 24 college credit hours.

I moved to Tucson and attended the University of Arizona on the GI Bill. In 1952, I was commissioned as a second lieutenant in the Military Police Corps, U. S. Army Reserves. In 1956, I competed in the Olympic trials for the USA weightlifting team. I placed third, but only the first two places went to the Olympics in Melbourne, Australia.

I was offered a regular commission as a captain on the U. S. Army weightlifting team that was training in Hawaii. This weight lifting team was the best at that time. I did not accept the regular Army commission. Instead, I became a real-estate developer, and with my wife and three children, continued my civilian career in Tucson, Arizona. In 1960, I was discharged from the Army reserves. My life would never have been so successful had I not made the enlistment in the regular Army as an underage volunteer.

Paul Ash lives in Tucson, Arizona.

# David Baillie, M2509

*Age 15 – United States Army*

I was born in the Lower East Side of New York City, New York, on 30 November 1934. We lived in a cold water walk-up on the fifth floor. My bassinet was the lower drawer in a dresser. My father was an artist and my mother a nurse. My father was from Scotland and had served in the United Kingdom Army for over 20 years, including service during World War I.

I left high school in the 10th grade and decided to join the Army. With a little help from an Army recruiter and my mother, all the papers were signed so my date of birth was "right" and I joined the Army at age 15. This was in December 1949.

When I enlisted I wanted to go into ordnance, but there were no openings. My second choice was engineers, but again no openings. I asked what was left and was told infantry or armor. I chose armor.

I took the oath at 1 Whitehall Street, New York City, New York, and spent the night on Governors Island before going to Fort Dix, New Jersey. There we all got shots and worked our butts off for a week learning how to march, drill, and fire and care for our weapons. God, were my arms sore from the shots! They felt like lead for days.

We were shipped south by train and I took basic with the 30th Infantry Regiment (training), 3rd Infantry Division (Audie Murphy's old division). Then I went to advanced training and took special courses in armor and infantry tactics, mine-laying, etc. I fired expert with the M-I rifle and other weapons. With the recommendation from one of the platoon sergeants, I was made an assistant instructor at the infantry school. We did all types of field problems with National Guard troops and later with students in the Basic Officers Infantry Course.

The word came down they were looking for men to go to Japan as replacements with the 24th Infantry Division. The 24th had been on occupation duty in Japan since the close of World War II. It was said there would be faster promotions and overseas pay if we went to Japan. So, I raised my hand and by April 1950 I was on my way to Japan after a I5-day leave home to Long Island, New York, to say goodbye to my mother (who had been a single parent since I was two months old).

We landed at Yokohama and were stationed nearby. I got a few 3-day passes to go to Tokyo, expecting it to be all bombed out, but it was not. It was a very beautiful city. By now I was a corporal. With some more training, we got ready to relieve the David Baillie men of the 24th. I was assigned to the 34th Infantry Regiment.

Then all hell broke out. North Korea invaded South Korea and my regiment was ordered to Korea, arriving there in early July. We landed with about 1,980 men and by August there were only 183 left standing. I had been assigned to the 34th Tank Company as an assistant section sergeant. All the tanks in the company had been hit or destroyed. I had three tanks knocked out from under me. Thank God the North Koreans were lousy shots. They would hit the treads only and disable the tanks. So we became infantry more times than not!

All the United Nations forces were pushed back south to Pusan and it was touch and go for some time. We got replacements and held the Pusan Perimeter. By now I was a platoon sergeant. There were no other non-commissioned officers (NCOs) left in the company. The 34th was made a paper unit and the flags and what was left of Headquarters Company went back to Japan. The rest of us were sent to other units.

I was in seven of the ten campaigns in the Korean War and stayed in the Far East Theater for some years. I was made a sergeant first class just before my last tour in Korea. I did a few tours with the 1st Cavalry Division in reconnaissance on the demilitarized zone (DMZ).

I went back to the States and was sent to noncommissioned officers school and later to the NCO Academy. I returned to Japan and was assigned to various posts in the Far East such as Indo China and Vietnam. I did several tours in Vietnam, my last with the 1st Cavalry. By now I was a first sergeant (E-8). By the time I left Vietnam I was on the list for sergeant major. Once I was back in the States again, I was promoted.

I left active duty and went into the Army Reserve for 10 years and retired in 1982 as a command sergeant major with a total of 33 1/2 years of service.

While I was still in the Army, I took the GED test and passed, so when I came back to the States I went to college on the GI Bill and became an educator for about 20 years. I am a published author of several books related to my Army service and college years. In 1999, I received the Stud Terkle award from the State of Illinois.

In between all this and after Vietnam, I was married and had 8 kids, all grown now. I have 21 grandchildren. I retired from teaching in 1999 because of health. I now do volunteer work with the Boy Scouts of America, the American Legion, VFW and AMVETS, where I am a district and national officer. My wife says that I'm not retired!

*David Baillie lives in Leicester, North Carolina.*

# Richard Baldwin

*Age 14 - United States Army*

My name is Richard Baldwin. I was born December 13, 1934. I was born in a little place right outside Sissonville, West Virginia, called "Coon Creek". I went to a one room schoolhouse. I did not have a birth certificate or anything like that. My Mother had a midwife and not a doctor. So when I got to be about 14 years old, my mom and dad were divorced. I was afraid that my mom could not afford to take care of me and my little brother. She was working a lot of hours for almost nothing. I just did not think she would be able to keep up. So I decided that I would go into the Army and take some of the pressure off of her.

I enlisted into the Army in October 1949. They needed a birth certificate or verification. So I got a woman to sign the papers and say she was my mother and vouch for my age. She listed that I was 17. When I first went into the Army I went to Fort Knox, Kentucky for Basic training. Those ten and 15 mile hikes were hard on a 14 year old. But I knew if I fell out that I would never live it down. After that I went to Fort Mead, Maryland in B Battery of the 39th Anti Aircraft outfit. We went to Camp Perry, Ohio for firing range over Lake Erie. While we were there the Korean War started. They sent me to Camp Breckenridge, Kentucky after that. This is where they confronted me about my age. So after I was in there two years; they sent me home. I believe someone told on me. That was September, 1951. They gave me an Honorable Discharge and I headed home.

When I got home, I was still too young to do much good at a good job. I was able to join the Teamsters Union when I was 16 years old. However, I did not tell them I was 16. They assigned me to the loading docks until August 1952. I was able to work out of there a little while and made some pretty decent money before deciding to join the Marines. My Mother had to verify that my age was correct and this time I got a Birth Certificate with my correct age on it. Then I waited until I turned 17 to go into the Marines.

I stayed in the Marine Corps for three years. They sent me to Paris Island, South Carolina for recruit training, then on to Quantico Marine Base, Virginia. I requested overseas duty. I thought I was on the way to

Korea. I was sent to Camp Pendleton, CA. Here at Camp Pendleton I was assigned to the 3rd Marine Division.

We were sent overseas from there. However, during the trip, a cease fire was called. The whole 3rd Marine Division was on hold in Japan to wait and see how the cease fire would go. They told us we could be on the front lines in 12 hours if need be. But they never had to move us. While I was in Japan we had maneuvers over in Okinawa.

We did a lot of good sight seeing there. While over there we went over to Iwo Jima. On Iwo Jima we reenacted the WWII battle. Some of us ran up Mount Suribachi. And some of the guys raised the flag there like the guys did in WWII. There was a lot of scenery there that I will never forget. I was in Japan about 15 months.

We came back through Fort Ord, California, then to Quantico and worked on the Potomac Bay running a landing craft. I also had extra duty running a cabin cruiser. I transported people across the Bay to the Indian Head Army Base on the Maryland side. When I came out of the service that time I was about 20 years old in August 1955. I went back to the Teamsters, and I was able to get me a better job. Soon, I got married and settled down. I was with the Teamsters 30 years. I went into sales in 1981. Everything has gone along fine in my life. I am still working in sales and plan to retire the end of this year. I am now 74 years old, soon to turn 75 in December. I don't think I ever regretted anything that has been done.

## IN APPRECIATION FOR THOSE WHO DID NOT MAKE IT BACK

Jack D. Harris, Minister

As a tiny infant, he was wrapped in blue
The joy he brought was love made new,
His parents offered God their thanks
As their new son joined the human ranks,
Through his childhood years, he quickly grew
And was taught to live by standards true.

His boyhood came with tee shirts and jeans
And he found his friends among Christian teens,
He was privileged to live in the land of the free
Surrounded by a loving family,
His faith helped him learn what life means
That spiritual truth outweighs material things.

His manhood saw him dressed in olive green
As proudly he chose the military scene,
A faithful soldier in a band of brothers
He was committed to serve for the sake of others,
Although his choice made short his dream
His life left a legacy for the rest of his team.

As a child of God, he's now clothed in white
In God's presence, his faith has become sight,
The arms of angels lifted him from his buddies' hands
And the wonders of Heaven replaced Iraqi's sands,
At a young age, this veteran faced his death
But in eternity, he took his first breath!

Be it civilian or soldier, when a young one is killed
Bitter tears flow as emotions yield,
But when one is taken in the line of Christian duty
Death's ugliness fades in the light of God's beauty,
The Lamb of God has conquered the power of the grave
His Life He freely gave, that our lives, He may save!

Composed April 19,2005
(For all who were killed in time of war)

# Kinchen D. Barker, LM1239

*Age 15 - United States Army*

I was born in Franklin, Venango County, Pennsylvania on 26 July 1926. I had two brothers and four sisters. We were a very close family. When I was 1 year old, my family moved to Porter's Hollow, West Virginia. This is where I was raised until I was 15 years old.

In early 1941 they drafted my oldest brother into the service, supposedly for one year. It broke my heart. He was seven years my senior. He spent time with me and taught me many things. He was my idol, my hero.

When the Japanese bombed Pearl Harbor on 7 December 1941 and war was declared the next day, I knew that my brother would have to serve for the duration of the war, plus six months. I missed him so much that I wanted to be with him. In my mind, I wanted to protect him. I asked my parents for permission to join the service, but, of course, they wouldn't give it. I threatened to run away from home. My dad knew that I would run away, so rather than take the chance I'd get into trouble or get myself killed someplace, he consented to sign my papers.

The first of March 1942, we went to the recruiting office and filled out the paperwork. I was 15 at the time. A few days later, I left for Fort Thomas, Kentucky, where I was given a physical and sworn in on 11 March 1942. After getting all the processing done at Fort Thomas, we all shipped out by train to Camp Cooke, California, which is now Vandenberg Air Force Base. This was a brand-new camp. We joined a brand-new division, the 5th Armored. We trained at Camp Cooke, then went on maneuvers in the Mojave desert near Needles, California. In November when maneuvers were over, we went back to Camp Cooke and a lot of us went home on furlough.

During the first few weeks that I was in the Army, we were asked if anybody could play a trumpet. I had started to learn the trumpet and knew a little bit about it, so I raised my hand and they automatically made me the company bugler. I had never played the bugle before, but it didn't take me long to learn the calls. In those days, every company had its own bugler.

By the time we were on maneuvers, we didn't have buglers any more. We didn't have time for them.

In March 1943, we went to Tennessee for maneuvers, then went by train to Pine Camp, which is now Fort Drum, in New York. We trained at Pine Camp for a few months and then went to Indiantown Gap, Pennsylvania. After training there for a few months, we went to New Jersey by train to board the USS D. B. Alexander, a troop transport which had 5,000 men aboard. On our way across the Atlantic we were in convoy. Our ship was in the center of the convoy, and as far as the eye could see in any direction, there was nothing but ships. I think that was the largest convoy that ever crossed the Atlantic.

We landed in Liverpool, England, and went by train to Swindon,

*The 5th Armored Division lead the drive toward Falaise, France, which we always referred to as Falaise Gap.*

England. We trained and maneuvered in the Salisbury Plain for the invasion of Europe. We crossed the English Channel and landed at Utah Beach on 26 July 1944. Our division was part of Patton's Third Army, but we were on the secret list and were never publicly listed during the war. We were known as Patton's Ghosts.

The 5th Armored Division lead the drive toward Falaise, France, which we always referred to as Falaise Gap. We surrounded the German Seventh Army and pretty much wiped them out. From there we moved on towards Paris. We could have liberated Paris, but we were instructed to stand still and let the French Second Division have the honor of liberating their own capitol.

We were the first American division to reach the Seine River. We went on to help liberate Belgium. We liberated Luxembourg and we were the first American division to fight on German soil. We went through the Siegfried Line and across the German heartland. We were 45 miles from Berlin when we were stopped, not by the Germans, but by orders form Supreme Headquarters Allied Expeditionary Force (SHEAF).

Right after the Germans surrendered, my outfit was put in a little town called Houngandran, Germany. It was in the Russian Zone of Occupation, but we were there temporarily until the Russians came in to take over. We stayed there for about a month.

Because I was five points short for discharge, I was sent back to the States, along with several others. After a short furlough, we were to go San Luis Obispo, California, for further transport to the Far East in preparation for the invasion of Japan. Fortunately for me, the war ended while I was on furlough. Instead of going to California, I went to Fort Meade, Maryland, was discharged and went home. This was in July 1945.

While I was overseas, I wrote to my mother and told her that when I was discharged, I was going to go back to school. I did, and I started in the ninth grade. After about three months, the principal called me into the office and told me that I was too far advanced to stay in high school. He said that there was something new that had come out. There was a test that I could take to get my diploma. I took the test and got my high school diploma, which made my mother very proud.

In1946, I re-enlisted in the Army and in October I was sent to Korea. I was assigned to a construction outfit that was to help rebuild the country. I was told to locate an area where a well could be drilled to produce water for a laundry to serve 5,000 men. I located an area and the well produced enough water for a 10,000 man laundry.

I returned to the States in March 1948 and was discharged again. After the North Koreans invaded South Korea in June 1950, I went back into the Army. They didn't send me to Korea. They sent me to Fort Knox, Kentucky, to train troops. I guess with my war experience and the experience I had in Korea, I was a logical choice to help train the men who would go to there.

I was sitting in the barracks one December evening when a guy came in. I thought I recognized him as one of the troops I had trained. He came up to me and said, "Sergeant Barker, I came by to thank you for what you did for me, for training me. It saved my life. I came by to thank you for it." You can imagine how that made me feel. I can't explain the feeling that came over me when that boy told me that. It made everything worthwhile for me.

When my time was up, I took another discharge. I was just like a yo-yo, going back and forth. In 1957, I joined the Army for the last time. I made up my mind that I was going to stay in and retire, which I did. I was sent to Heidelberg, Germany, and spent 4 years and 10 months there. When I returned to the States, I was stationed at Fort McNair, near Washington, D.C. After some time there, I went to Fort Huachuca, Arizona, in 1967. Two years later, I was transferred to Fort Monmouth, New Jersey. I returned to Fort Huachuca and retired there on 1 January 1972. My active duty and reserve time totaled 28 years, 8 months, and 13 days. I enjoyed my time in the service.

As a civilian, I was a jack-of-all-trades. I drove truck, I built houses and fences, and did about anything that needed to be done. I worked in Florida for about 16 years and moved to Tennessee in 1990.

Dave Barker and his wife Elizabeth Joanne live in Crossville, Tennessee. They have a combined family of 5 children, 8 grandchildren, and 6 great-grandchildren.

# Theodore Bishop

*Age 16 - U. S. Army*

January 22, 1949 I was 16 years old, when I enlisted in the army. I took my basic training at Fort Jackson South Carolina. After basic training I reported to Fort Lewis, Washington to the Quarter Master Corp, the 529th Petroleum Supply Compo We provided gas to all of Fort Lewis.

In 1950, we shipped to Korea and arrived on August 4th. We were in what they called "The Pusan Perimeter"; I was 17 years old then.

Later in early November 1950, I was transferred to the 2nd 10 9th regiment "c" company.

When the Chinese came into North Korea the 2nd 10 had to retreat. Lt. Jackson told me to get behind the wheel of a 2.5 ton truck that didn't run. All I could do was to steer the truck. It was freezing and a truck in front of us was towing us. The mud from the truck ahead of me was splashing on my wind shield. Lt. Jackson had to use the bayonet to chip the ice and mud off. I don't remember where or how far south we went.

While in North Korea in combat I got frost bite on both feet. I was sent to the hospital in Osaka, Japan and I was treated for it. While in the hospital I came down with Hepatitis and was sent to another hospital in Kyoto, Japan to be treated. Then they sent me back to the hospital in Osaka, Japan.

Soon I was sent back to my old unit, the 9th regiment; the "C" company. We got into a fire fight and we were running low on ammo. I volunteered to go to the rear to get more ammo. I got about 20 or 25 Bandalers and two machine gun barrels.

I could hardly walk with all the ammo and I only weighed 150lbs, but I did get back in. One thing I'll never forget was the bloated bodies in the rivers and streams. We used to laugh about it, eating our "c" rations sitting next to the bloated bodies ... but where else could we go?

My Sgt. in Korea was Sal Cantu. He was also a young warrior. We stay in touch with each other.

I served with my unit until August 4th 1951. I was awarded the "Purple Heart Combat Infantry badge", Korea Service medal with 3 battle stars and

United Nation Ribbon.

I came back to the states at Fort Benny, Georgia where I served 3 years, 4 months, and 18 days.

After being discharged I went to work for the phone company (S.Bel!) as a cable splicer helper.

I retired after 31 years as an assistant manager.

I am now a member of the American Legion, the O.A.V. I serve as Bugler in the American Legion Honor Guard. I also do some volunteer work at the local VA hospital.

# Hank A. Bourque Sr.

*Age 16 – U.S. Navy*

I was born Dec. 13th 1954 to Opha and Lois Bourque one of seven children from my fathers second marriage, and three brothers from his first marriage.

I grew up in a small town by the name of Jennings, LA. My father was a cab driver for the town of Jennings and my mother was a homemaker.

We lived on the southside of town which wasn't the best area to raise a family. The streets was a place where you fought or you got picked on so I decided to fight.

The people I hung around with was trouble so I began to move away from this crowd. At school I was above average in all my subjects and wanted to further my education. I completed the 10th grade.

I had two of my older brothers serving tours in Vietnam and felt it was time to enlist. I spoke with my father he stated he would sign if my mother would give the OK. I spoke with my mother who refused to sign. I continued to ask my mother to sign the documents and finally she stated she would sign. But no Marines or Army so I chose the Navy. I was sworn in Oct. 13, 1970.

I reported for basic training at Orlando, FLA. After basic I was assigned aboard the U.S.S. Savannah AOR-4 as a gunner. We were informed the ship was headed for deployment to Vietnam. Before the ship left, I was transferred to Cuba for more training. Our training consisted of hand to hand combat, repelling from helicopters, and hands on sprinkler systems.

I really wanted to be assigned to Vietnam but my ship was deployed to the Mediterranean Sea for six months. I never got to go to Vietnam. That was my only regret for the years I served in the navy.

I am proud to have served my country during the Vietnam War. I earned numerous medals and citations. Then on April 24th, 1974 I was honorably discharged.

I married the sweetest country girl in 1973. We have 4 children and five grandsons.

I would like to thank my older brother Joe Bourque featured in the Young Warrior's Magazine in Feb-March Issue for inspiring me to serve our great country.

Although we never shared our lives together as a family we do have love for each other. We live within three miles of each other and see each other often.

# Ophelias Joseph "Joe" Bourque, M2562

*Age 14 – United States Marine Corps*

On 23 July 1942, my life journey began some 180 years after the arrival of the French Acadians in Louisiana. My parents were both descendants of the original families who were exiled from Nova Scotia following the invasion by British forces. The Bourque family had settled in the southwest part of Louisiana in the town of Gueydan. My birth marked me as son number three in this Cajun family.

When I was only two weeks old, my father abandoned my mother and his three sons, leaving us in a very difficult position. Since our country was locked in the throes of World War II, times in our small town were hard. Consequently, my mother was faced with an awesome responsibility. There were no governmental assistance programs available from which we could gain help. Jobs were scarce, but Mom was a woman who was determined to survive and to find a way to care for her boys. Finally, because she was an excellent cook, she secured employment in a local restaurant and for a time she kept us together.

Mom wanted more for us than she was able to provide, so she sought a way for us to have a better life. She decided to place her three sons in the care of the St. Mary Orphanage in Lafayette, Louisiana. I was only two years old when this decision was made. Although my mother had the very best of intentions when she chose to do this, it did not prove to be the best for us.

Since my first two years had been filled with such tension, the loss of my father and the poverty of our family, I had some problems. The most obvious was that from time to time I was not able to keep from wetting my bed as I slept. Even though I could not help doing this, I was severely punished. When the nuns came to check our beds in the morning, if they found that my bed was wet, I was whipped with a bamboo switch. Each time this happened, my back, buttocks and legs were covered with large red welts. My middle brother, who was four years older than I, slept in the same dormitory where I slept. When he saw the beatings that I received, he decided he would do something about it. He would awaken before me and check my bed. If he found it to be wet, he would move me from my bed to his and then take my place in the wet bed. He received the beatings that would have been mine. Although I was quite young at the time, I will never forget the sacrifice he made for me.

We lived in the orphanage for five years, and after this we were able to be reunited with our mother. Our time under the strictness and the rigid schedule the nuns subjected us to served to embitter us and fill our minds with anger. The three of us learned how to survive in the streets. We were often in fights with anyone who suggested any kind of challenge. We would take whatever we could wherever we found it, regardless of the consequences. As I look back on this formative period, I realize that we were always in trouble and were probably responsible for much heartache in our dear mother's life.

By the time I was 14, I had become a juvenile delinquent and was getting deeper and deeper into trouble. I knew that it was only a matter of time until I was arrested and sentenced for my bad behavior. I took my middle brother's birth certificate and hitchhiked to Houston, Texas. When I

---

*I have been privileged to live the American dream, and I am honored that I could proudly wear the uniform of an American soldier for twelve years.*

---

arrived there I knew no one; I had no money, no place to go, and I was very hungry.

Even though I was only 14, I was taller than most 14-year-old boys, and I looked older than I actually was. I was informed that if a person joined the military, he would be given food and lodging, so I decided to give it a try. I went to the Marine Corps recruiting officer and told him that I wanted to enlist. I was accepted at once. I had to take the physical and the written tests, but I passed each of them easily. I was sworn in early the next morning and put on a plane for San Diego, California, and was taken to the Marine Corps Recruit Depot. For the next several weeks we were guided through the tough exercises that would turn us into Marines. I actually enjoyed being given this opportunity. We graduated in November 1956, and I was given orders to report for duty at Camp Pendleton, California.

As much as I was enjoying this structured life, it was to be short-lived. One day I was ordered to report to the company commander. When I entered his office he told me that he knew that I was only 14 years of age, and because of this, I had to be released from the Marines. After I realized that my masquerade was over, I confessed, was discharged and sent back to my mother in Louisiana.

When I arrived back home, I was forced to return to school. I began my classes and continued them just long enough for me to reach age 16 when I could legally quit school. Two months after I walked away from the classroom I landed in some serious trouble, was arrested and was brought before the juvenile judge. He sentenced me to serve two years in the Louisiana State Reformatory for my crimes. I was taken the Louisiana Training Institute for Boys in Monroe, Louisiana. I fully expected the absolute worst when I arrived there, but I soon discovered that I was badly mistaken.

The people who made up the staff at the institute were decent and caring people who really wanted to help redirect the wayward boys who were sent there. Without any hesitation, they showed their goodness and their love for all of us who were incarcerated. We were required to attend school each week day morning and then, in the afternoon, we were given different kinds of responsibilities. Quite often, I was privileged to drive the truck whenever it became necessary.

One of the men who really had a positive impact on my life was our coach, Mr. Fred Lindsay. He was instrumental in bringing a meaningful change in my outlook on life. With the constant encouragement from Mr. Lindsay, for the first time in my life, I really wanted to do better, to rise above the level on which I had been living. I still feel a deep indebtedness to this great man and to the judge who sentenced me.

I completed my sentence at the institute in September 1960 and decided to return to the military. I enlisted in the U. S. Army and was sent to Fort Riley, Kansas, for my basic training. After advanced infantry training, I became a member of the 1st Cavalry Division and was deployed to South Korea where I served for 13 months. I returned to Fort Leonard Wood, Missouri, and became a drill sergeant. Eighteen months later, I returned to South Korea, this time with the 7th Infantry Division, where I served for 16 months. I was in "The Land of the Midnight Calm" for a total of 29 months. I returned to Fort Leonard Wood again as a drill sergeant.

In October 1965, I was ordered to report to A Company, 1st Battalion, 35th Infantry Regiment of the 25th Division, in Hawaii. Two months later my outfit was deployed to the Republic of South Vietnam, arriving there on 3 January 1966. Shortly after our arrival, we were sent to a spot near Pleiku, where we were to construct a fire-base camp. The 25th Division was assigned search and destroy missions throughout the Pie Me area. On one mission in an encounter with the enemy, I took refuge behind a large tree. A rocket or a mortar round struck the tree and the resulting explosion caused me to lose my hearing. It was three days before I could hear again. In a later engagement, I was struck in the thigh with a piece of shrapnel. I was evacuated to the 6th Convalescent Hospital where I remained for a month.

After my release I returned to my unit and was given orders to set up an ambush at the fork of a river, and to sink every vessel that we saw. After three days without sighting the enemy, but fighting with millions of mosquitoes, we returned to base. Eight of us were taken with malaria, necessitating a 14-day confinement in the hospital. Our days in South Vietnam were anything but pleasant, so I was extremely happy when my tour of duty there ended.

Upon my return to the States, I was assigned to Fort Lewis, Washington, to help train new recruits for combat. I was determined to do the best job I could to train these men who would soon be risking their lives in Vietnam.

While at Fort Lewis, I received my GED from my old high school in Kaplan, Louisiana. In 1973, I received a medical discharge and returned to civilian life. I enrolled in McNeese State University in Lake Charles, Louisiana, majoring in education. After receiving a BA degree in elementary education, I received a position as a classroom teacher. In 1994, I received a MA degree in administration and supervision. I retired in 2005.

I have been privileged to live the American dream, and I am honored that I could proudly wear the uniform of an American soldier for twelve years.

Joe Bourque lives in Jennings, Louisiana.

# C. T. Bowman

*Age 15 - U.S. Navy*

My trip started out like many other veterans of underage military service. The year was 1944. The war was still raging on both sides of the oceans. I felt an obligation to serve my country. The only problem was that I was only 15 years of age. I needed to find someway to join the Navy.

I spoke with my father and was happy that he agreed to sign stating I was 18 years of age. I went down to the Navy recruiting office and requested permission to join the navy.

I passed all of the tests except one. I was very tall for my age but, I was very skinny. The recruiter told me to go home, eat a lot of bananas and drink plenty of water. The next day I weighted in and passed.

My first assignment was to the Great Lakes Naval Station for boot camp. I did just fine and felt real good about myself. Upon completion of boot camp I was assigned to the U.S. Naval Station in San Diego for additional training.

After training I was assigned to the USS Cavalier. Our main job was to ferry combat troops to the war zone. We would then load the wounded to be transferred back to the U. S. for medical care.

Next I received orders transferring me to the USS Goodrich DDE31. Our ship left San Diego headed to the Atlantic. We traveled through the Panama Canal. This was a sight to see. Our ship continued to Key West, Florida where we pulled into port. We were allowed to go ashore for a brief period of time.

Our ship continued up the coast to Newport, R.I. This would be our home port. Next we received orders for two goodwill trips to Greece, Italy, and France. Most of these areas were stilled damaged by the war. We were allowed to visit each port. We had a good time.

In Feb. 1949 at the age of 19 I received an honorable discharge from the U. S. Navy.

After my discharge I began training at the navy yard in Washington, D.C.

I am retired now. My wife Velma and I have 4 children and 4 grand children. We reside in Palm Bay, Florida.

# J. Clyde Broome, LM1999

*Age 15 - U. S. National Guard*

J. Clyde Broome, Colonel AUS Retired Veterans of Underage Military Service LM1999. My first knowledge of the military came on December 7, 1941 when I heard about the Pearl Harbor attack. Even at nine years old, I knew that something bad had happened I went out on to the front porch of my home and shouted "To hell with the damn Japs" and immediately was pulled back into the house! The local airbase was expanded and troops began to arrive in town. Four railroad companies serviced my town (Columbus, Miss.), and

I saw many trains carrying personnel and equipment. All of this activity caused me to be interested in what was happening military wise. Then in 1948, my best friend joined the local National Guard unit and encouraged me to do likewise. I told him that I was only 15 and probably could not get in. He said that recruiters did not ask him for a birth certificate and probably would not ask me, either, and they didn't. I told them that my birthday was July 17, 1930, not 1932. This was enough for me to pass. In December 1950, the local high school was out for Christmas break when the alert came down that our unit would be mobilized in January. The principal called in the seven of us, who were seniors and in the Guard, and told us that we would receive credit for that semester and would be given military credit for the second semester, enabling us to be able to receive our high school diploma's. It turned out that when May came around, the Commanding General of the 31 St Division issued five-day leave for all graduating seniors, high school, junior college, and college, to return and graduate with their class. One requirement was that we had to graduate in uniform. When the seven of us walked across the stage to receive our diploma's, the community could then see how it had been impacted by the callup.

In January, my unit, Battery A, 932 Field Artillery Battalion 31st Infantry Division, was mobilized and sent to Fort Jackson, SC for training. I was a Sergeant and Chief of Section of a 105MM Howitzer. Since only World War II veterans had basic training, we had to give ourselves basic and advanced training. The division received about 5000 draftees from mainly the Northeast and the Midwest. This influx of personnel was culture shock for all! After we had completed all of our training, many were being sent to Korea as replacements. I was promoted to Sergeant First Class, and after completing Field Artillery Leadership Course, was sent to Korea in September 1951 where I became a Chief of Section for a 105MM Howitzer in Battery A, 57th Field Artillery of the 7th Infantry Division. As we entered into winter, there was not much movement of the front lines. We were above the 38th parallel and in the mountains where it was very cold. As our supported regiment went into reserve, my battalion would relocate and reinforced the fires of other battalions, which were still on line. An artillery battalion is given one Final Protective Fire mission, which is last resort fires for the infantry. When the battery is not firing other missions, the data for the FPF is set on the howitzers so as to not delay firing when called upon. This is one mission that you hope never comes up.

However, on Thanksgiving Eve 1951, we received the command to fire the FPF, which means continuous firing. It was called the barrage, in those days. We fired all night long as fast as we could. Ammunition became a short supply, but the ammunition trains were able to keep bringing rounds to us. The next morning, the attack was over, and when we cleaned the tube, I noticed that there was not one land that did not have a break in it. The breaks were from two inches to two feet long. This was the result of continuous fire during the night. I pulled my howitzer out of position and went to ordnance where a new tube was replaced. After my tour in Korea, I returned to the United States and rejoined the Mississippi National Guard where I received a field commission to 2nd lieutenant. The criteria was that, if you had been a First Three Grader and held a leadership position in combat for more than six months, you could be promoted to that rank. During the next two years, I worked at the post office and got married. I decided that I wanted to make a career of the Army and applied for active duty. There was no program for recall of officers, so I resigned and enlisted as a Master Sergeant. I was sent to Fort Bragg where I ended up as a Battalion Operation Sergeant in the 266th Armored Field Artillery Battalion.

Shortly thereafter, I was on my way back to Korea for my second tour. I arrived in November 1954 and was assigned as the Battalion Operation Sergeant of the 48th Field Artillery Battalion, 7th Infantry Division. The 48th Field had a Direct Support Mission of a Marine rifle company located on three Ops on the DMZ. The Marines were relieved by the 24th Infantry Division and due to some rearrangement of the mission, and due to the shortage of officers, I ended up in the 52nd Field Artillery Battalion, 24th

Division and was assigned as the Liaison Officer to the 21st Infantry Regimental. After about six months, we started receiving officers, and I was reassigned as the battalion operation sergeant until I rotated in March 1956. Little did I know that the knowledge and experience that I gained in these assignments would be very helpful as my career continued. After returning to the States, I was discharged from active duty and recommissioned in the National Guard. After a couple of years, I took advantage of the GI Bill and later was graduated from Mississippi State University. By this time, we had a 4 year-old daughter, and I was seven years behind my college peers when I went to work. My first job was in New Orleans, La., where I continued my military with the Louisiana National Guard. Later, we returned to Columbus where I had worked during college and rejoined the Mississippi National Guard. I was assigned as the battery commander of a divisional artillery headquarters. After reorganization, the unit was converted to an Honest John Rocket battalion where I commanded Battery A, which had 8" towed howitzers!

My company transferred me to North Carolina in 1965, and I joined the North Carolina National Guard where I had several interesting and challenging jobs. When I retired in 1988, I had served in all artillery officer positions with the exception of Sin and Disease, chaplain and surgeon! During this period, as many of you know who were connected to the National Guard, we went through many reorganizations. On the way, I served as the S-3 and S-1 of a field artillery group, assistant S-3 and S-2 of a division artillery headquarters, executive officer and commanding officer of a 8" self-propelled artillery battalion, S-3, executive officer, and commanding officer of a field artillery brigade. One of the highlights of my career was to be selected and graduated from the U.S. Army War College. My last three years were served as the mobilization designee at the state headquarters where I was the Nuclear Surety Officer and director of the no-notice mobilization team, which exercised the units in preparation for active duty. In1988, after almost 40 years of service, I retired as a Colonel. If I had to do it over, I would do it again; however, some of my decisions over that period would not necessarily be the same!

# Clinton W. Bushacker

*Age 16 – Army Aircorp*

Born in Danbury, TX, Oct 28, 1926. Moved to Austin, TX for my 5th birthday. My brother (who was a VUMS) was home from Ft. Sam Houston for the weekend and we listened to the Jap attack on Pearl Harbor. The following year I enlisted in the Army Air Corp for my 16th birthday and reported to Dodd Field, TX (next to Ft. Sam) Nov. 8, 1942. I was shipped to Bowman Field, Ky. to complete my basic training and assigned to the 55th Troop Carrier Squadren 375th T.C.

I was sent to the Douglas Craft Co. in Long Beach, CA for my aircraft maintenance training. We deployed to Lorenburg, SC for further training and then to Ft. Wayne, ID. Next in to Camp Stoneman, CA for shipment to SWPA via Australia in to New Guinea, May 1943.

I served in New Guinea as an aircraft mechanic and flight engineer on C47 and a Cargo Drop Version of the B17 Boeing Bomber. In 1943 my brother was reassigned from the 2nd Division Engineer to the 183 Division Engineers for deployment to England in 1944. While they were being trained and prepared for the invasion of France my mother decided she didn't need two sons in the war as I was only 17 then (this was about March 1944). After about three months the Army Air Corps decided to ship me back to the states for Honorable Discharge.

After all the travel and processing I was finally discharged June 9, to head home when the European Invasion was started. I went to work for the Army Air Corp at Bergstrom Base as an aircraft mechanic. Although I was 17 then the draft board wouldn't leave me alone. So then I joined the Navy and shipped out to San Diego Naval Station on Sept. 5 1944. After basic training again I was sent to Norman, OK Naval Base for training in Aviation. I was reassigned to Edotin Naval Air Station, NC. Although I volunteered for overseas several times I finished the war at Edotin, NC. I was in the air force reserves and was recalled to active duty July, 1952 for the Korean conflict. I finished my career working as a civilian logistic specialist with 42 1/2 years total service when I retired at age 57.

# Robert G. "Bob" Button, LM2409

*Age 16 - United States Navy*

About five years ago I was sworn into an aviation flotilla of the United States Coast Guard Auxiliary as a pilot. I fly my personal aircraft on patrols and search missions for the Coast Guard. When I crank up and that big prop turns, I'm a member of the U. S. Coast Guard, same as anyone else who wears the Semper Paratus uniform.

The fact is I've been wearing my country's uniforms for more than 60 years, starting when I joined the Navy in 1945. I had turned 16 in July that year. The Navy might never have known 1 was underage had I not got sick. In sick bay I was delirious and couldn't for the life of me remember my phony birthday. The next day a chaplain showed up at my bedside, and in March 1946, I was honorably discharged. The discharge papers read: "Underage in authorized enlistment," authorized because I had lied about my age but hadn't forged any documents.

In September 1945 the local draft board set up shop in the gym at Union Hill High School, Union City, New Jersey. I was playing hooky when I went to the see the draft officials, so 1 had to sneak around hoping that none of my teachers would spot me. I signed up for the draft using Ordinary Seaman's papers that said I was 18.

I had gotten the Seaman's papers a year earlier when I conned my way into the Merchant Marine. I never did get to ship out on a merchant ship though, because I looked like I was 12 years old and nobody wanted a "baby" on their crew. We'd gather inside a union hall near Battery Park in New York City. All the old salts would look at me and laugh. I'll never know how I later fooled the draft board.

I had made it through boot camp at Bainbridge, Maryland, and had been assigned to a ship when I got sick. The Navy sent me home, but it was fun, except for that first night in barracks -1 was so scared I wet the bed!

Joining the Coast Guard Auxiliary as a pilot at age 72 still gives me a chuckle. During those intervening years I served in every branch of service - Navy, Air Force, Army, and Marines. Donning that Coast Guard uniform made it a clean sweep: five branches - six if you count the Merchant Marine - 1 don't, because I never served on a merchant ship. Yes, 1 do get a tad weary trying to pass those physicals at my age. I'm a bit long in the tooth for that rescue swim test, for instance. But it has been a wonderful

30-year combined active and reserve military service, not to mention the few decades I've logged as a civilian since 1945.

I joined the Army Air Forces shortly after my Navy discharge in 1946 and served in a night-fighter squadron of P-61 Black Widows, a deadly World War II twin engine fighter. Before my hitch ended, the Army Air Forces became the United States Air Force, so my discharge reads USAF. I tried civilian life for a while, and went to college on the World War II GI Bill.

When the Korean War broke out I hungered to get into combat, so I re-upped and went to the 2nd Infantry Division's 23rd Regiment in Korea. Well, as they say, be careful what you wish for - Heartbreak Ridge, Old Baldy, Pork Chop Hill - battles that mean nothing to most folks nowadays are etched into my memory forever. It was a bloody damned war, aptly called the "Forgotten War."

Before Korea I was in the 505th Parachute Infantry Regiment, 82nd Airborne Division, where I got some of the best infantry training anyone could hope for, and it served me well in combat. When I got home from Korea, I was assigned to Baker Company, 3rd "Old Guard" Infantry Regiment, the ceremonial unit at Arlington National Cemetery, the oldest continuously active military unit in existence. I'm afraid I tend to brag once in awhile over drinks about having served in such elite outfits.

I got out of the Army in 1953, but because I had been a specialist in long-range recon patrols behind enemy lines in Korea, the Marine Corps recruited me to travel the country and teach escape and evasion. The President's Code of Conduct had just been enacted. Every soul in uniform had to be taught how to behave in the event of capture. So, as a Marine buck sergeant I became an instructor.

The Army Special Forces was organizing reserve units around the country, recruiting airborne veterans with combat experience, so after I got out of the Marine Corps in 1960 I signed up as sergeant major of a 21-man B team and donned the green beret. I was once again a civilian, writing documentary films for a living in San Diego, California.

I also worked on several newspapers as a reporter, but soon was activated again by the Army Reserve as second in command of the First U. S. Army Leadership School in Pennsylvania. Once that brief hitch was over, I went back into the Reserve as a senior Public Affairs NCO, finally retiring from the Army as a sergeant major (E-9) in 1989.

I've been flying airplanes for more than 40 years, which is how come the Coast Guard approached me to join their auxiliary at the tender age of 72. That was five years ago and I'm still doing my best to get out of the sack and into the air for Uncle Sam every chance I get.

Bob Button lives in Jersey City, New Jersey.

# WM BURNHAM CARTER -LM1865

*Age 16 - U.S. Marine Corp*

I was born in Savannah, Ga., on May 25, 1933. I was second of 6 children in the family of a divorced mom.

The family had a history of alcoholism, un-employment and lack of support for dependent children.

The most responsible role model was a paternal uncle Wade Carter who was killed in action in Essen, Germany during WWII. Although deceased Wade Carter continued to be an inspiration to me. I determined that I would serve my country in the military as he had—even at the risk to my own life. I was 8 years old when WWII began; and 12 years old when it concluded. Due to my-circumstances, I started my first job after school at 11 years old.

I was on the "waiting list" to be admitted to Bethesda Orphanage when I joined Junior ROTC at age 15 in the sophomore year of high school.

On November 3, 1949, at age 16 I enlisted in the USMC Reserve Unit in Savannah—without parental permission.

I was 17 years old when we were activated and went to Camp Pendelton—although due to my age I could have obtained a discharge—I refused.

While I was undergoing training, older members of our unit were sent to Korea where they were being killed & wounded. I had to attain my 18th birthday before I could go to Korea. I arrived in Korea in July, 1951 and was assigned to the 5th regiment USMC as a squad leader. I saw action in the "punch bowl" East-Central area. I was advised we were subjected to the heaviest mortar-artillery attacks by the enemy. I was slightly wounded; did not get recognition because I was assisting the corpsman with a fatally wounded forward observer.

I completed my tour; and returned to the U.S.

When I returned; I had completed my GED although I had missed the last two years of high school. I subsequently finished four years of college in three years; earned my Masters & Ph.D. degrees and became a college professor and psychologist.

I subsequently served as Captain in the "Georgia State Defense Force". I sincerely believe that I fulfilled my commitment to my country, U.S. Marine Corps and my uncle Wade Carter—RIP.

# James Cooper LM1954

*Age 16 - US Army*

As I attempt this writing, my thoughts are of my current and past friends of the army as well as my current and future children.

I was born near Blytheville, Arkansas area in Jan 18, 1933. Not much there in the 30s. The only thing we had in abundance was "Hard Times." This was the peak of the great depression and it did not look like it was going to get better. By location, we lived near the Mississippi River. As time went by our area had its share of military members as the draft was in full swing and many locals were enlisting. To a young kid that had been nowhere or had very much, the military started to look very good.

Early on, I started think that the military was the place for me and chance to escape the area. My mother had died when I was age four due to cancer. Dad was a farmer and was fortunate to marry a wonderful lady that was like an angel to us. Dad was able to get a bigger farm about 600 acres near Helena, AR and this was our next move. Unfortunately dad was working as a farmer or drinking and this caused a divorce and we lost one great step-mother. Soon he married again, this time she was not an angel, she moved in and brought with her 7 kids. She did not want me, my five brothers. Early on at the age of 12, I had experience on the farm where we raised cotton, hay, com, soy beans and yes, we still had the victory garden as most people did. In addition to this we had horses, milk cows, beef cattle, and hogs. Now, we have everything but a mother and money.

At age 12 with pressure from the second step-mother, I left home. By age 14, I was qualified to do the work of most men. By 1949 like a lot of other youngsters (under-age maybe 50 or more) in Arkansas, it was time for me to get in the Army or give it up as a bad day dream. I chose to enlist at age 16 and my Army service started at Camp Chaffee, Arkansas. After graduation from basic training, I was shipped to Seattle, WA and that was a big trip and far from the State of Arkansas. Soon, I was shipped to Japan.

While stationed in Japan, the Korean War started with North Korea invading South Korea and this would be my next move. We were the first U.S. Division to arrive Korea, 4 July 50. Here, in Korea I was in the 24th division and we saw our full share of the war. Our General Dean was captured and served as a POW.

The country had few to no hard surface roads. Mostly ox cart trails that better suited jeeps than six by sixes. Over the years, I would return to S. Korea for two additional tours (1957-8) and (1964-5). Changes between last two tours were unreal. It went from small mud hut villages to large multistory city buildings. Today, S Korea is one of the economic countries of the world including the manufacture of fine automobiles and more.

I completed my enlistment as a sergeant at age 18 now and I took to leave back home to see my family. However, second step-mother was still there and the only thing that changed was she had more kids. This became only a visit to me and was just the encouragement that I needed to re-enlist with the Army. Times were still very bad in the 1950s. Upon my re-enlistment, I was sent to Europe. By now, my mind was made up that this was the place for me.

As my thoughts go forward in time, it is now the 60s era and time for another war. My next war assignment was in the country of Vietnam, in history you see that it was known as Indo China. In 1950 French were at war with them. Here in Vietnam, I served with the US Army 19th infantry. The days and nights can be very long and the food not good, even the water was bad in a place like this and chances for injury is all around. Speaking of water, I did not know what a rain storm was until I spent some time there. I was assigned to the Vietnamese Army Infantry as a Battalion Advisor in Mytho-Mekong Delta in 1968, Earlyon, I was commissioned as a second lieutenant. Also, I was wounded and by the year of 1968, I was retired for permanent combat disability as a Major at age 36.

In retrospect about war in general, "No Normal Soldier" likes war. But the sad truth is that SOME wars do help some of those involved. For example, South Korea would never have become the educated economic power in the world today had the US never stepped in to rescue them.

My most happy days in the Army started with my enlistment and this was followed with many good times and some very bad times. Did you ever notice that in life we tend to recall the good times quicker than the bad times. I told a friend about the time in

France, we were not allowed to frequent certain bars. However, a couple of NCOs that were regulars at this made it look alright to me. They invited me along and while there, I went to the rest room and upon my return, my friends were gone. I asked the French Bar Maid, where did they go? She said, the MPs come and take away! Wow, was that good timing for me. My friends did not reveal that I was their guest on that one and only visit with them at the off limits place. However, for me, it was good lesson that I will never forget.

I retired and told no one that I went in at age sixteen. About 30 years after my retirement, I learned that a lot of kids did the same thing by tacking on an extra year or so to their age. There were so many that did so that an organization known as Veterans of Underage Military Service, (VUMS) was formed by Alan Stover a former US Coast Guards Man that enlisted at age 14. He did extensive research and was able to get letters from each branch of service declaring that no problems would be encountered by those that enlisted underage. Now, in the year of 2009 much history has been complied about them and it is believed that nearly 250,000 enlisted from WW II through Vietnam. One of our members known as Frank Buckles enlisted into the army during WW I and at the time of this writing (2009) it is believed that he is the only survivor world-wide from the First World War. He celebrated his 108th birthday in February of 2009.

Further, I have heard that some of the boys and girls were in at age 12,13,14,15, and 16. The girls could not join until age 20 and before that many of them had graduated from high school. Several of the women are members of VUMS. Some of the boys were prisoner of war at age 14 and on. One of the boys earned the Medal of Honor in while serving in WW II. There have been 3 books compiled to date by the title of America's Youngest Warriors by Editors Drs Ray and Susan Jackson. (Also, members of VUMS) Each book has 199 stories in it and a beautiful cover by one of our members. You may find the books in libraries in most state capitols. As I read those books, I do so with a sense of pride and yes, I would do it all over again with the same existing conditions.

Jim and his wife Christa of 56 years reside in Albuquerque, NM Their family includes two daughters and two grand-children. Jim is a member of the American legion.

# William Daniels

*Age 15, U.S. Navy*

"Well, I enlisted in Birmingham (Ala.) at age 15, and stayed through world War II. I tried to enlist about half-a-dozen times before I made it," Daniels said. "You get a little smarter every time they throw you out. Finally, I got smart and went with a family Bible to prove how old I was." Daniels had obtained the Bible from a book store, filling up the blank pages with a fanciful family "history." It did the trick. After a physical, he was sworn in. Daniels' first wartime sea duty was assignment to a provision store ship, the USS Calamares (AF 18) as a seaman in 1943. The ship departed for the Admiralty Islands right away, "a long haul," Daniels recalled. Down through the Soloman island chain and the north coast of New Guinea. His next move was to the port director's office, visual communicator section signal tower on the island of Biak, off the north coast of New Guinea, as an signalman communicator. "I worked there with the Australians for quite a while. Several months later, when Gen. Douglas McArthur invaded the Philippines in October 1944, Daniels was moved to Tacloban, on the island of Leyte, in support of the invasion. When the war was over, Daniels didn't return to the United States until February 1946. "I think everybody in the United States forgot we just had a war by the time I got home," he joked. Troops rotated home according to a point system, and the only thing he had points for was "time out of the states." Which in his case didn't amount to much. Unlike many service members with families—due to his age — he didn't have any dependents. Daniels stayed in the Naval Reserve, going on to the Naval Academy in 1950 and graduating in 1954. The underage enlistment caught up with him at the Naval Academy, when he was called to the commandants office to account for the discrepancy in his records. The JAG officer at the academy, a Pearl Harbor veteran who was missing a leg, must have taken the younger veterans wartime service into account. "We'll get everything straightened out," he told the commandant. Two weeks later, everything was in order. As an ensign, his next assignment was sea duty on celebrated destroyer USS Laffey (DD 724). Laffey was a combat veteran, nicknamed "the ship that would not die," after withstanding kamikaze attacks off Okinawa. After duty aboard Laffey, he went on to submarine school in Groton, Conn. Daniels served aboard three

different submarines: all diesel-powered boats — Sea Cat (SS 399). He made two Arctic Ocean cruises while in submarines, engaging in cable-laying, patrols and other activities. "They were a couple of cold winters, from 1957 to 1958."

But with the dawn of the nuclear powered submarine, an entirely new education process was called for in training. "Hyman Rickover decided I was too old for his nuclear program when I graduated from school." I was 26 when I got out - so I told my detailer, "you better get me a dadgum job, because there isn't going to be a diesel submarine in about an hour and a half around here," smiled Daniels.

Not long afterward, he found himself aboard USS Princeton (CV 37, LPH 5), an Essex-class aircraft carrier hauling Marines and helicopters into Vietnam starting in 1965.

"I made a total of 17 amphibious assaults off of Princeton," Daniels said. "We would haul a Marine reinforced battalion and two Marine helicopter squadrons (at a time)." As first lieutenant on Princeton, all the deck divisions were his onboard charge.

Daniels received orders to become the commanding officer of a Naval Reserve center in Omaha, Neb. next, for about a year. Then he was off to Washington, D.C., where he began a relationship with the Navy's Bureau of Personnel (BUPERS) that would be a lasting one. Though he wasn't there long, he'd later return to BUPERS.

Daniels returned to sea as executive officer of USS Ponce (LPD 15), an amphibious transport. From there he went directly to command of DD 862, USS Vogelgesang, a Gearingclass destroyer. He kept command of Vogelgesang for about two years, returning to BUPERS again after that.

His next assignment was to command a destroyer squadron, DesRon 34 homeported out of Charleston, S.C. "There were five ships in it, six before I finished up," he noted. "I enjoyed that tour very much."

Though he returned to BUPERS again, for a total of four tours, he became deputy commander of Naval Reserve Surface Force in New Orleans, La., after that, retiring there at age 57 in 1985 as a two-star admiral.

Reflecting on the turns his career had taken over the years, "I enjoyed every bit of it," the once-underage Sailor admitted.

# Homero A. del Castillo,

*Age 15 - United States Army*

I was born in Laredo, Texas, on 17 August 1932. Jobs were tight and we were staying with our widowed grandmother, Rufina Garibay. My grandfather, Zacharias Alvarez del Castillo, was a lieutenant colonel in the Army of Mexico, and was killed in combat during the battle of Zacatecas.

My brother Alfred and I would sell lemons house-to-house on Saturdays and Sundays. We charged a nickel for nine lemons. In the early 1940s, we sold newspapers. In 1944, I started working at a fruit stand in downtown Laredo. My salary was $3.50 per week. The salary seemed low; however, there were perks in my job. My boss had three restaurants and I could eat lunch and supper at any of them.

The war against Germany, Italy, and Japan was being fought. The bus stations and downtown Laredo were crowded with soldiers, sailors, Marines and Air Force personnel. Fort McIntosh was a small Army post and the nearby Laredo Gunnery School was training hundreds of soldiers in the art of shooting machine guns aboard aircraft.

In 1945, I was delivering the Laredo Times to the El Cuatro neighborhood where I was born. This area is close to downtown and is now called the Laredo Historical District. While attending school I could see the soldiers from Fort McIntosh Homero del Castillo marching on Victoria Street. I felt very proud of my country and I wanted to be a soldier real bad. In high school, I attended ROTC classes and enjoyed learning about military responsibility.

One morning I was suddenly embarrassed by my English teacher in front of the class. She told me, "You are unable to learn the parts of speech." Students who were better educated than I started giggling and laughing. For a few moments I remained silent. I then looked straight at her and replied, "The parts of speech of which I am not able to learn does not mean I will not succeed. I have already learned one of the ways to succeed since the third grade without knowing the parts of speech. Abraham Lincoln learned from reading and I will follow his example."

My parents were informed of the incident and the teacher's behavior. This incident left a bitter taste in my mouth and I decided to part company and look for an educational alternative.

Many of my friends were enlisting in the military and the few of us who were younger were being left behind. Some of them obtained blank baptismal forms and began filling in the blanks by hand. I could see right away that this was an obvious weakness that the recruiting officers would spot right away. On my own, I patiently developed an acceptable document. The day finally came when I took several tests and was scheduled to ship out to Fort Sam Houston, Texas. I spoke with my parents and told them I was going to enlist in the U. S. Army. The Army presented to me a great opportunity to better myself.

On 5 April 1948 I was sworn in at Fort Sam Houston and shipped to Fort Ord, California, by train for basic training. I was assigned to B Battery, 29th Field Artillery Battalion, 4th

*The war against Germany, Italy, and Japan was being fought. The bus stations and downtown Laredo were crowded with soldiers, sailors, Marines and Air Force personnel.*

Infantry Division. Army life was all I thought it would be. My weight was 127 pounds and I was 5 feet 6 inches tall. My stature was small, but I put in a full day's work. I was very motivated and very happy.

Soon after completing basic training I was sent to Camp Leroy Johnson, Louisiana, with a few days' delay en route in Laredo, Texas, to visit my family. Upon my arrival at Camp Johnson, I made it a point to visit New Orleans as often as possible before we shipped out to Panama.

We boarded the troopship Private William H Thomas to begin our voyage to Panama on 26 July 1948. On 1 August I was put on detail to clean one of the ship's offices and empty two butt cans. I climbed a steel staircase and balanced myself with a handrail on my right and carried the two butt cans in my left hand. When I reached the top of the staircase, I took a step with my left foot and hit two soldiers who were passed out on the steel platform which was the entrance to the deck above, blocking my entrance. I lost my balance and fell backwards down the staircase. There was no way to avoid hitting my head, back, shoulders, arms and knees several times until I hit the steel deck at the bottom. I lay there feeling severe pain all over my body. Some time later I began making attempts to get up and finally I was able to make my way to my bed.

The next morning we had arrived in Panama. I put on my class A uniform and found my left foot swollen. I could not put my shoe on. A soldier took my duffel bag up on deck for me. I was standing with some men with whom I had taken basic training with when a captain appeared. He asked me why I was out of uniform. I told him about my accidental fall. The captain took me to the ship's physician who examined me after I briefed him about my fall. I was put on a stretcher and taken by ambulance to the hospital at Fort Clayton. At the hospital, I briefed the doctor about my fall. A cast was put on my left foot, but none of the other injuries I had told the captain and the ship's physician about were examined, treated, or documented.

As the days passed I experienced more and more irrational feelings. I was discharged from the hospital and was taken to B Battery, 903rd AAAAW Battalion at Fort Clayton. Due to irrational behavior and amnesia caused by my head injury, I had problems with a master sergeant while at the hospital, and later while on guard duty I incurred the wrath of the officer of the guard. These problems continued, but I was never referred to a military doctor. I was sent back to the States and was given an honorable discharge on 16 November 1948.

On 18 August 1949 I reenlisted in the Army and served at Fort Lewis, Washington, with the Quartermaster Service Company. After several months our unit was transferred to Alaska where I served for 30 months.

Upon my discharge I returned to school, but my injuries from the fall aboard ship continued to plague me. My grades suffered and I was not able to continue. I feel that because my injuries during the fall were not treated nor documented, I have had much difficulty in getting medical help from the Veterans Administration (VA). At one time the VA awarded me a service-connected disability, then reversed it ten days later, saying, "We made a mistake." All this time my psychiatric problems continued. I was able to obtain some help from the University of Texas medical branch at Galveston. I still feel that my injuries were not treated by Army doctors nor the Veterans Administration doctors.

Homero A. del Castillo lives in Laredo, Texas.

## THE PROFILE OF A HERO

By Jack D. Harris

I recall that day in August of Nineteen Forty four,
When word came to our neighbors' front door,
"We regret to inform you," the telegram read
Suddenly they knew their soldier son was dead.

"Killed in action," he died a brave man
On a nameless hill, in a far off land,
He did his duty, the best he could
He'd stood his ground, as they knew he would.

I'll never forget that piercing scream
It still brings sadness, like a bad dream,
A friend of our family was forever gone
How quickly it seemed, his life had flown.

For years his father had worked on the railroad line
As a man of God, he was gentle and kind,
He'd raised his family to be their best
In their deep sorrow, God gave them rest.

His son entered the army from East Tennessee
And hoped to help bring a quick victory,
When this soldier died, he left a young child
A daughter with memories that forever she filed.

Before his death that day in France
He'd met a Texas Lad on his way, by chance,
An orphan boy, only five foot five
As a father and son, they'd stayed alive.

"Brandon's" death brought pain to this Texan's heart
His blood gave courage an unstoppable start,
With fire in his heart, since "Brandon" had died
The boy became a man, with all fear aside.

His natural skills sprang into effect
He rallied his men and led them to connect
With the valor of a warrior, born in madness,
He led his company to forgo their sadness.

When the war was over and the Nazi nightmare
Ended in disgrace and had come to despair,
This young Texan who'd loved "Brandon" so much
Was received with the honor of a nation's touch.

More than half a century has entered time's tomb
On thousands of graves, new flowers now bloom,
"To Hell and back," is the story they told
The legacy they gave is worth more than gold.

"Brandon's" life was short, but his memory remains
Etched in the fabric, that our history retains,
Let us never forget the sacrifice these made
Let us always remember the price these paid.

Let us join and thank God that we live in a time
Though troubles abound and our streets know crime,
In spite of these things, and because of great men
Our nation is worth defending again.

(Written as a tribute to Pfc. Lattie Tipton, a native of Erwin, Tennessee, affectionately known as "Brandon," until his death following D Day in World War II, and who figured significantly in the life of our nation's most decorated soldier, Lt. Audie L. Murphy.)

# Ernest H. Diesslin, LM1565

*Age 16, United States Army*

On 6 August 1946, I forged documents to show that I was 17 years old. I needed a parent's signature and my mother reluctantly signed when I told her I would get in with or without her signature. I went to Fort Snelling, Minnesota, which was about ten miles from our home. I was there for about ten days. I was not issued a uniform and I was allowed to go home every night and return the following morning.

Eventually, I was sent by train to Fort Riley,

Kansas, where I was issued a uniform, etc. and from there, I was sent to Camp Polk, Louisiana, which is now Fort Polk. I had about four weeks of basic training. On the rifle range, I qualified as expert rifleman with the M1 Garand rifle. Marching and 20-mile hikes were the order of the day. I was then sent by troop train to Kansas City, Missouri. Due to a delay in route, I was able to return home on a 10-day furlough. Upon returning to Kansas City, we were sent by train to Camp Stoneman, California. After a wait of 10 days, we boarded a troopship headed for the Pacific, eventually arriving in Japan.

I was assigned to a malaria control detachment.

We did a lot of spraying with DDT mixed with kerosene. We also used ground meat laced with antu, a deadly poison, to eradicate rats. We wrapped it like candy kisses and placed them in lots of five in corners and against inside walls in military storage buildings. After about a week, we would check, counting the remaining ones. This would continue until all the bait was gone.

During the year I spent in Japan, I learned to speak Japanese. When my 18-month enlistment was nearly up, I was sent to Fort Lawton, Washington, for discharge. I returned home to Minnesota.

In May 1949, I joined the Navy for three years.

My enlistment was involuntarily extended for a year due to the Korean War. I was injured while I was in the Navy and received a 10% disability. When my four years were up, I left the Navy. During those years I completed a GED for high school and also one year of college. Looking back at my military service, my best experience was in Japan.

Ernest Diesslin lives in Holden, Massachusetts.

# Duke V. Doering, LM2619

*Age 15 - Army National Guard*

After Lieutenant General Robert Foley, the 6 foot 7 inch Commander of Fifth U.S. Army, signed my Legion of Merit award, I reflected back on my choice to join the military at the early age of 15.

Sometimes it is not the things you did that you regret, it is the things you did not do. Joining is one of those things I am very thankful that I did. When I was 15 I really had no idea what I wanted to do.

As it turned out it was a fantastic experience. The one thing about joining and working in the military is that I found all soldiers, regardless of what era they served, had a great loyalty to the United States of America. After I completed the oath of enlistment, which started with raising my right hand followed by the words "repeat after me", I was sure that enlistment was the proper thing to do. I recently read this quote from an unknown author, "A veteran - whether active duty, retired, or reserve - is someone who, at one point in their life, wrote a blank check made payable to 'The Government of the United States of America' for an amount of 'up to and including my life'. That hits the nail on the head. In reviewing the jobs I held, the places I traveled and most of all, the people I met and worked with, I think my enlistment was a very good choice.

It all started in the year 1955 when the principal of the Sturgis, SO High School was William J. Brown. He was a combat veteran of World War II, a Commander of an engineer company in North Africa and Italy. Then in 1950 he was recalled as a Lieutenant Colonel and commander of an engineer battalion that was activated and sent to Germany during the Korean War. Brown was a soldier, a patriot, an educator, a great American and strong authority figure. When Mr. Brown called me, a sophomore in high school, into his office in September 1955 it was to be a routine counseling session, and it started that way. He was surprised that I had a grade of A in English Literature and a B in typing. He seemed concerned with my grade of C in wood shop. His comment was that boys usually get better grades in wood shop than in literature. I started to reply but he cut me off. That was the short and abrupt end of the discussion of my grades. He then went on to ask me if I had ever considered joining the National Guard? I replied that I had not, I was only 15. His reply was "why don't you go down to the National Guard armory after school and talk to Warrant

Officer Murphey, tell him that you are interested in learning about the Guard and maybe enlisting". So I did. After school I walked into the basement of the armory and told him that I had talked to Mr. Brown and I might like to join the National Guard. Murphey asked "how old are you? I replied that I was 15. To that he said "you have to be 17 to join the Guard, and that is with your parents consent". I said OK and started to leave when he said, "why don't you walk around the block and come back and see me?" This made absolutely no sense to me, but I was terrified of authority figures, which Mr. Murphey was. I went outside and started walking around the block, all the time thinking, what is this all about? In about 10 minutes I arrived back where I started. I walked into Mr. Murphey's office and he looked at me like he had never seen me before, then he said, "what can I do for you young man?", suddenly, out of the blue it struck me and I said, "I would like to join the National Guard!". He then said "how old are you? and I said "17, sir". He said "good, take these permission papers home to your parents for their signature and return them to me". On September 15, 1955, at age 15, I was sworn into the unit.

National Guard meetings were held for 2 hours every Monday night with two weeks of annual training, which was at that time called "summer camp". My first summer camp was when I had turned 16 years old and was assigned as a truck driver, driving a M-211 2 1/2 ton dump truck (deuce and a half). I did not have a drivers license, either civilian or military, but nobody asked or seemed to care. I drove carefully and with various assistant drivers, who always seemed to be a crusty old sergeant.

I was assigned as the Construction Draftsman, MOS 811.10, for the unit. In my senior year I received an A in drafting. So on one fine June morning I was shipped off to Ft. Belvoir, Virginia for the U.S. Army Engineer Schools 10 week construction draftsman course. About halfway through the course it got to the point where I just hated drafting. A little bit was good, but setting at a drafting table for 8 hours a day was driving me nuts. Although I did quite well academically in this Army course, I never spent another minute at a drafting table after graduation from that class. I was glad that I had good typing skills as it seemed everyone in the army needed a typist, and that was something I liked better than drafting.

On July 5,1963 I was hired on "full-time" as the Administrative-Supply Technician (AST) for Headquarters Company, 109th Engineer Battalion. Subsequently in 1968 I transferred to the 842nd Engineer Equipment Company. I was a Platoon Sergeant of 1st Equipment Platoon. During the period 1963 through 1975 I was sent TDY to Ft. Carson, Colorado; Ft. Riley, KS; Ft. Ben Harrison, IN; Ft. Leonard Wood, MO; Ft. MacArthur, CA; and Ft. Lewis, WA for various schools, training and missions.

I was later transferred to the Headquarters. My transfer to Headquarters became effective in 1976 as a Master Sergeant. I served in that rank until I was appointed Warrant Officer, WO 1 on July 4, 1978. It was later that year that I was sent TDY to the Pentagon to assist the publication staff on writing and editing Army Regulation 220-1 and Department of the Army Pamphlet 220-1. 6th U.S. Army at the Presidio of San Francisco was our immediate headquarters in the 1970s and 1980s. I was fortunate enough to be assigned TDY to that Headquarters on several occasions. It was at that location that I completed the Force Integrators Course that was presented by the staff at the Command and General Staff Instructors from Ft. Leavenworth, KS.

Later I was periodically assigned TDY to the instructor staff at Ft. Leavenworth for force development and force structure classes.

In 1989 I was TDY at Fort Clayton, Panama when the country was under control of Manuel Noriega. Tensions were tight between the United States and Panama. I had just returned from a mission in Honduras, deplaned a cargo plane at Howard AFB, and got back to the fort just in time to witness a demonstration at the front gate of Fort Clayton by the local dissidents. It was painfully obvious to me that the problems between the two countries were not going to be resolved easily or peacefully. It was in December of that year that the United States military implemented mission Just Cause, and removed Noriega from control.

Competitive marksmanship was not only a valuable skill, but was also a pleasant diversion from the normal routine. Many happy hours were spent firing match grade .45 caliber pistols at places like Ft. Ord, California; Ft. Riley, Kansas; Ft. Lewis, Washington; Camp Robinson, Arkansas; Nashville, Tennessee, Camp Guernsey, Wyoming; Camp Santiago, Puerto Rico and many civilian regional matches around the country. Being awarded the U.S. Army Distinguished Pistol Shot Award from Ft. Benning, GA was a culmination of that effort.

When the Army opened a new Warrant Officer rank, Chief Warrant Officer 5, I was fortunate enough to be in a position to complete the Warrant Officer Senior Course at Ft. Rucker, Alabama and be pinned as CW5. I retired as Chief Warrant Officer 5.

I had graduated, early in my career, I was awarded a Bachelor of Science Degree with majors in Business Administration and Accounting.

My wife, Susan, and I have two daughters who live in Florida and the two sons who live in South Dakota. We have five grandchildren.

# Robert E. Dykes, LM2553

Age 15 – United States Air Force

I was born near the small town of Jonesborough in Washington County, Tennessee in 1940. Jonesborough is the oldest town in Tennessee and served as the Capitol of the State of Franklin, which later became the State of Tennessee.

I was the third child of twelve children. My Mom loved to say that she had six of one and a half dozen of the other - six boys and six girls.

Eventually, all six boys would enlist in the military. Larry, the oldest, entered the Navy at age 17. Later on Albert joined the Navy. Carl enlisted in the Air Force and then did a hitch in the Navy. John chose the Marines, and Raymond picked the Army. When Larry left for the Navy, I figured that I was next to leave. I could hardly wait. After completing the ninth grade, I was employed at an automotive body shop. I really liked the job, and according to the shop owner, I was learning the trade very quickly. However, because of a persistent truant officer, I was forced to give up my job. In February of 1956, I went to the Air Force recruiting office. I passed the entrance examination. After getting my parent's signature, I was off to Knoxville, Tennessee for a physical. After passing the physical, our group of 13 enlistees was sworn in on February 14, 1956. I was fifteen years old and on my way to San Antonio, Texas.

I spent eleven weeks in basic training at Lackland Air Force Base (AFB). I reported to my first duty assignment at Craig AFB at Selma, Alabama on 17 May 1956. There I was placed in on the job training (OJT) for the aircraft machinist career field (531xO).

I served five years at Craig where I was exposed to the racial problems that plagued the south in the 50s and 60s. It was hard for me to understand why there was so much hatred between the races. I had two African-American friends at Craig, both of whom I had gone through basic training with. We arrived in Selma by train but were not allowed to share a taxi to the base. We were not allowed to socialize off base because of segregation practices in effect at that time.

I left Craig in May of 1961 for an 18-month tour at Kadena Air Base, Okinawa.

Leaving there in October of 1962, I was assigned to McGuire AFB,

New Jersey. En route to the United States from the Pacific, the pilot briefed the passengers concerning President Kennedy's speech regarding the Cuban Missile Crisis. From McGuire, I was next assigned to Ubon Royal Thai Air Base, Thailand on 14 January 1966. I met my brother Carl in California. He was returning from a 12-month tour at Bien Hoa Air Base, South Vietnam.

Twelve months later I returned to the states where I served at Tyndall AFB, Florida, until I left for Vietnam in August of 1968. My two Vietnam assignments were both at Phan Rang Air Base, South Vietnam. The two tours were separated by a 13-month assignment at Kessler AFB, Mississippi. After completing my second tour at Phan Rang in November 1971, I reported to Westover AFB, Massachusetts.

After only a few months at Westover, the entire wing was sent on temporary duty (TDY) to Anderson AB, Guam and Kadena AB, Okinawa.

---

*Unlike many of the World War II veterans who received a short training session, then went directly to the front lines, I spent 10 years on active duty before serving in a war zone.*

---

The B-52s went to Guam. I deployed to Kadena with the KC-135 tankers. At the end of the 6 months TDY, I returned to Westover only to find out that the base was closing. I was reassigned to Shaw AFB, South Carolina, where I retired on 1 March 1976 with 20 years service. I was thirty five years old.

Unlike many of the World War II veterans who received a short training session, then went directly to the front lines, I spent 10 years on active duty before serving in a war zone. I have great respect for all veterans but I especially have a lot of respect and admiration for the World War II veterans who were there for the duration. I agree with Colonel Ollie North, "Their war story is a story that deserves to be told".

One highlight of my career came about while stationed at Ubon AB, Thailand. My wing commander was Colonel Robin Olds. He planned a "dogfight" mission over North Vietnam in early January 1967. His plan involved disguising the F-4 aircraft so that the North Vietnamese would think it was an F-105, which was a much slower aircraft. Part of the deception was to hang an F-I05 pylon on the F-4. The pylon had to be modified to fit the F-4. That is where I got involved. I was the Non-

Commissioned Officer in Charge (NCOIC) of the machine shop where the pylons were machined to fit the mount on the F- 4.

As a result of Colonel Olds' plan, The North Vietnamese took the bait. The MIGs came up after what they thought was the slow-moving F-105. When they came up through the clouds, they got the surprise of the day - a sky full of the far superior F -4s. Our wing shot down seven MIGs that day. I believe Colonel Olds got the first one.

While in the military, I made friends with some remarkable people. It is impossible to mention them all. There is not a day goes by that I do not think of them. I would like to mention a few in particular.

While stationed at Craig AFB in Selma, Alabama I met Marvin (Mac) McKenzie who was from Louisa, Kentucky. Mac worked on the aircraft flight controls. He and I became great friends. Mac presently lives in Milford, Ohio. We stay in touch.

During my first tour at Phan Rang AB, South Vietnam, I met Bobby Parton. Bob was from White Pine, Tennessee not far from Jonesborough. He was also an aircraft machinist. Bob and I made many trips together throughout South Viet Nam on aircraft recovery missions. He presently lives in Azle, Texas. We visit often.

While serving my second tour at Phan Rang Air Base, I met Hiram Earl Gossett.

He was an aircraft machinist. Earl and I became friends and have kept in contact through the years. Earl and his wife Judy live in Griffin, Georgia.

After finishing his Navy enlistment, My brother Carl returned to the Air Foree and served in the Air Force Reserves until he retired in 1995. Carl and his wife Jeanette live in Trussville, Alabama.

After retiring, I returned to Jonesborough. I had received my GED while in the Air Force, so I decided to continue my education at East Tennessee State University. In May of 1980, I graduated with a B.S. degree in nursing. On 31 December 2002, I retired from the Veterans Administration Medical Center, Mountain Home, Tennessee. I was serving as Nursing Supervisor at that time. My wife and I enjoy traveling. We frequently stay at military camp grounds and temporary lodging when available. When at home, I try to help my pastor at the church we attend. There is always work to be done. Thank you for letting me share my story. I am proud to be a young warrior.

# Bobby S. Endfinger, LM2578

*Age 16 - United States Navy*

I was born in Pell City, Alabama, on 4 September 1927. Life was hard during the Great Depression, but I had a happy childhood.

My military training started on 25 August 1939 when I was 10 years of age. On that day, I went to the Alabama Boys Industrial School. It was the best military school in the world. The commander was Henry C. Woods, a World War I veteran. In 1940, I was slated to be the first lieutenant of Company F. I took Company F to the Veterans Day parade in Birmingham on 11 November 1940. All branches of the military drilled in the parade. My company won the award for the best company in the parade. We were awarded our medal in front of the Tutwiler Hotel in Birmingham.

Prior to the beginning of World War II on 7 December 1941, our school had 480 men and boys enrolled. Shortly after the Pearl Harbor attack, 300 left to go to war, leaving only 180 enrolled.

After receiving military training in this school, I was discharged to go back to public school. In 1942, I was asked if I would come to work at Avondale Mills to help out. Since so many men had been drafted and sent to war, they needed people to work so they could keep three shifts running the mill. The draft was taking so many men, and the mill needed people so badly, that they asked the school to send the boys who were not making good grades to replace the men that had left.

In September 1942, an employee of Avondale Mills said that if I wanted to go to work there I should see Dr. R. A. Martin. I did, and he gave me a birth certificate showing that I was born in 1925, which would make me 17 years old, although I was really only 15. So, I went to work for the Mills. I registered for the draft on 7 September 1943 using my new birth date and was drafted into the Navy on 21 October 1943.

My boot-camp training took place at Great Lakes, Illinois. I was 5 feet, 5 inches when I went in, and my seabag weighed 132 pounds. None of the clothes I received fit me. By the time I was out of the Navy I was 5 feet 7 inches tall.

One day we had to go to Camp Perry to perform. We marched 8 miles with our seabags on our shoulders. When we got there, we passed through a group of 200 men who had had to run until it got dark that day. Only five men were still on their feet. Those were the ones who had prior military training.

On 1 January 1944, we left for gunnery school in Norfolk, Virginia. We were to be assigned to a destroyer-escort. My name was not put on the payroll list at that time. I had to wash and press uniforms of others to have money for leave. Nine months later they found the pay list showing my name in Washington, D.C., but my name was marked through with red ink. By this time we were at the Navy ship yard in Charleston, South Carolina.

I was a member of the crew for the shakedown cruise of the USS

---

*Unlike many of the World War II veterans who received a short training session, then went directly to the front lines, I spent 10 years on active duty before serving in a war zone.*

---

Hodges (DE-231). When we returned to Charleston, South Carolina, we were all given a 7 -day leave. I could not go because I was sick and had no money. Finally, on 5 June 1944, I got paid for the first time. I had a 2-day pass, and for the first time I had enough money to go home to Pell City, Alabama. But getting back was a problem. I didn't get back in time, so I was AWOL. They put me in the Philadelphia, Pennsylvania, Navy Yard brig for three months.

Later that year I went to the South Pacific where I served aboard the USS Howard F. Clark (DE-533). I was a gunner on a 40mm antiaircraft gun. During the battle for Luzon Island in the Philippines, kamikaze planes hit our ships. After the battle, the Clark guarded the ship that provided military burials at sea for the men who had been killed. There was concern that Japanese submarines would attack during the ceremonies. I watched as they performed burials from sunrise to sunset, seven at a time from each side of the ship.

After the battle at Luzon, the Clark was sent to Iwo Jima. The battle that took place there was very difficult. Aircraft bombed the island so much it looked like the island was shaking. Our ship laid down a smoke screen and bombarded the island. Four days after the invasion, I could see the

American flag flying on Mount Suribachi.

After nine weeks near Iwo Jima, we went to Eniwetok Island to guard troopships. A Japanese submarine was torpedoing ships as they left the harbor. We located the submarine, dropped depth charges, and blew it up. We sailed for Okinawa to participate in the invasion. During that trip we learned that President Roosevelt had died.

At Okinawa, we guarded munition ships. After Okinawa was secured, we escorted supply ships and aircraft carriers back to Guam. While anchored at Guam, we were taking a break and watching a movie when a Japanese submarine blew up a tanker. This cut short our break.

The Clark was one of the first ships to search for the USS Indianapolis that was sunk after delivering the atomic bomb to Tinian. We were called back from the search to get ready to go to Japan. Near Japan, we picked up some of the crew from the American submarine Swordfish, which had sneaked into Tokyo harbor and blown up some ships anchored there. We were 64 miles offshore of Japan when the atomic bomb was exploded over Hiroshima. We could see the mushroom cloud.

After the war, we returned to the States. I stayed in the Navy until 12 August 1948. I returned to Pell City, settled down and raised a family of five children. I helped form VFW chapter 4758 in Pell City. For five years we were an All-American post.

Bobby Endfinger lives in Pell City, Alabama.

# Orley Fout

*US Navy, Age 16 in WW II and Recalled for Korean War 1950*

In retrospect to my early life prior to the US Navy, My mother had moved to Junction City, KS in June of 1944 with my 4 month old half

brother. My step dad had served a three year army enlistment starting about 1928. With the dawn of WW II, he was drafted in 1943 by the US Army and sent to Fort Riley, KS as a recruit instructor. In the mean time, all the boys a year or two older than me and the ones that I respected, either enlisted or were drafted into the service. I made up my mind to join the Navy, I went to city hall and got a copy of my birth certificate. My date of birth was July 21, 1928. I had to change the eight to a seven and this was accomplished by the purchase of a small bottle of ink eradicator, I removed the 8. However, my birth certificate was pale sky blue to white. I had to find a way to restore the blue color around the new number SEVEN. During my teen years, I used to play pool and while chalking up my pool que, I noticed the chalk was the same color of blue as my birth certificate. By using the pool que chalk, I rubbed and ground in the chalk over and over until the number seven looked like it was in a blue background. Then I crumpled the certificate and then smoothed it out several times until it looked like it had been carelessly stored for 16 years.

My next stop was the US Navy Recruiting Office where I presented my messy birth certificate and it was accepted. Next, the Navy sent my enlistment papers to Junction City, KS. I had already contacted my mom and alerted her that she would be getting enlistment papers to sign. She signed them upon receipt. On November 17, 1944, I traveled to Toledo, OH and underwent my physical. The doctor told me my ear drums were badly scarred and he would have my ears washed out with water under pressure. He told me to return home and come back in a week and he would look at my ear drums again. A week later upon return, I passed my physical exam.

Under orders by the US Navy, I went to Great Lakes Recruit training (Boot Camp) which started on December 8, 1944. Upon completion of Boot Camp, I was assigned to Basic Engineering School for 8 weeks. After completing Engineering School, I was sent to the

Naval Receiving Base called Camp Shoemaker, just outside of Oakland, CA. I boarded the USS Pickway, an APA Ship.

USS Pickaway (LPA-222) formerly known as the APA - 222 until 1969 Credit the Amphibious Photo Archive dept. (A brief on the USS Pickway as follows. Pickway was laid down by the Permanente Metals Corp., Richmond, California, as a victory cargo ship (MCV hull 570), on 1 September 1944; launched on 5 November 1944; sponsored by Mrs. William D. Schoning; converted into an APA; acquired by the Navy and

*Like many of my fellow sailors, I signed up for the inactive reserve. This meant no meetings, no training and no pay. The main benefit as an inactive reserve in time of a conflict would have the right to request active duty with the same rank that we had during the time of our release from the Navy*

commissioned on 12 December 1944, Commander. H. L. Bixby in command For those not familiar with the term APA, it is an Auxiliary Personnel Attack Ship).

We sailed on June 22, 1945 for sixteen days arriving at Noumea, New Caledonia on July 8, 1945. Noumea (French word) is the capital city of the French Territory of New Caledonia. It is located on a peninsula just south of New Caledonia's main island, Grande Terre, and is home to the majority of the island's European, Polynesian, Indonesian, and Vietnamese populations, as well as many Melanesians that work in one of the South Pacific's highly industrialized cities.

The arrival at Noumea, New Caledonia was 13 days before my 17th birthday. I was in New Caledonia less than a month. The consensus there was that an invasion of Japan would cost the allies at least 1,000,000 causalities. The ferocity of the Japanese on IWO and Okinawa foreshadowed the invasion of their homeland. We would have to kill almost every Japanese Citizen. I was shipped from New Caledonia to the island of Espiritu Santo, New Hebrides. Here, I was assigned to a Stevedore Company. We disembarked in the New Hebrides on August 9, 1945. We were so far off the beaten path that I don't remember anything about the second A Bomb that had been dropped that day on Nagasaki, Japan.

Early in the Fall, about October of 1945, I was in the recreation hall playing ping pong when I heard a voice ask, "Is Orley Fout present?" This was Navy Chief Petty Officer Hoover, he was of my acquaintance in New Caledonia. He ask if I would like to leave this ROCK and become a crew member on the APC 38. I think our complement including officers was less than 40 men. He said the electrician petty officer had left the ship at Guam as he had double the points needed for discharge. I left the new Hebrides on a ship that the main engine was out of commission. So we were under tow to New Caledonia, I believe it is about 500 miles or so. We went into dry-dock at New Caledonia. Upon completion of the needed repairs, we sailed for the US. We stopped at the American Somoa and again at the island of Palmyra for food and water. Our evaporator didn't have the capacity to freshen enough water for our needs. So, we stopped often to replenish our supply. The crew was not permitted to use fresh water to shower. We used salt water for our showers and then shifting it to fresh water just for a minute to rinse off. If we encountered a rain storm, almost all of the crew would stand on the deck in the rain water. This was like a bonus to our crew having been without fresh water to fill our needs.

The entire crew was made up of achieved shellbacks except a very young ensign that had been brought in by air craft to this area. He was warned of the ceremony to make him; a shellback. He was worried but since he was the only pollywog on board, we didn't initiate him. Upon arriving at the Port of Bremerton, I took a train home to Ohio for 25 days. I reported into the Naval Air Base at Gross Isle, Michigan. Then I flew to Great Lakes and was assigned to the Navy Pier for a few months, my discharge followed soon there-after on August 8, 1946.

Like many of my fellow sailors, I signed up for the inactive reserve. This meant no meetings, no training and no pay. The main benefit as an inactive reserve in time of a conflict would have the right to request active duty with the same rank that we had during the time of our release from the Navy. Also, the Navy reserved the right to recall us to active duty. I did receive a request from the US Navy to return to active duty for a period of six months. This was just three weeks after South Korea was invaded by North Korea. On June 25, 1950 I flew to the Great Lakes for a physical check up including shots and issue of a sea bag and clothes as needed.

Next, I reported to Norfolk, VA for assignment to the USS Missouri B63. The six months call to active duty was stretched to one year and again; I was discharged August 8, 1951.

In 1950, North Korea invaded South Korea, prompting the United States to intervene in the name of the United Nations. President Truman was caught off guard when the invasion struck, but quickly ordered U.S.

forces stationed in Japan into South Korea. Truman also sent U.S. based troops, tanks, fighter and bomber aircraft, and a strong naval force to Korea to support the Republic of Korea. As part of the naval mobilization Missouri was called up from the Atlantic Fleet and dispatched from Norfolk on 19 August to support UN forces on the Korean peninsula. The Missouri joined the UN just west of Kyushu on 14 September, where she became the flagship of Rear Admiral A. E. Smith. The first American battleship to reach Korean waters, she bombarded Samchok on 15 September 1950 in an attempt to divert troops and attention from the Inchon landings. This was the first time since World War II that Missouri had fired her guns in anger, and in company with the cruiser Helena and two destroyers, she helped prepare the way for the 8th Army offensive.

Missouri arrived at Inchon on 19 September, and on 10 October became flagship of Rear Admiral J. M. Higgins, commander, Cruiser Division 5 (CruDiv 5). She arrived at Sasebo on 14 October, where she became flagship of Vice Admiral A. D. Struble, Commander, 7th Fleet. After screening the aircraft carrier Valley Forge along the east coast of Korea, she conducted bombardment missions from 12 October to 26 October in the Chongjin and Tanchon areas, and at Wonsan where she again screened carriers eastward of Wonsan.

In concluding this story, I would do it all over again. I am very happy to learn that we have an organization called Veterans of Underage Military Service (VUMS) and that I can be a member and enjoy the stories that are close to my story.

Orley has spent his entire civilian life in the Marion, OR area. He is married to his wife Thelma or the past 56 years. They have 6 children and one of the boys served in Vietnam and was wounded severely on his last day. Also, have 14 grand-kids and 15 great-grand-kids. Orley is retired from the US Postal Service.

# Robert Joseph Hansberry, Sr.

*Age 16 - U.S. Navy*

One of America's Youngest Warriors, Robert was born on July 1, 1935, in Jersey City, New Jersey, and enlisted in the United States Navy, in September, 1951, at the age of 16 years, in Tampa, Florida, where the Recruiter, assured him that he would be able to complete his education in the Armed Forces, while fighting the nation's enemies. This proved to be true and Robert completed High School and two years of College while learning a profession and serving on an Attack Troop Transport (The USS Cavalier, APA-37), and an Attack Cargo Transport (The USS Skagit, AKA-lOS) in Korea and Vietnam.

Upon Discharge from the Navy, Robert enlisted in the United States Army, where he furthered his education, working on a Bachelor's degree in Mathematics, and continued serving his country in the Cold Wars and in Vietnam, where he served, first, as an electronics technician, and then as a Hawk Missile Technician.

While attending College, Robert became convinced that he had reached his personal limits in the military. He had married and sired four children, so he took a discharge, in 1962. He worked for the Department of the Army, at the Signal Long Lines Facility, in Davis, California, for three years, then transferred to the Department of Commerce, where he started at the Coast and Geodetic Survey, then for N.O.A.A., at the time known as The Weather Bureau. Both positions were at the Nevada Atomic Test Site, in Mercury, Nevada, and involved testing for adverse conditions due to Atomic Testing. In 1970, Robert learned of an opening at the Las Vegas International Airport, with the United States Federal Aviation Agency (FAA), applied for the job, and was transferred there until his first retirement. The airport was located very near the University of Nevada, at Las Vegas, and afforded Robert the opportunity to complete his education, by changing his major fields of study to Business with a Finance Minor, thus allowing him to use much of his Mathematical studies in the new field.

Upon retirement from Government Service, Robert went into the Sales Field, qualifying as a Real Estate Broker, and opened a brokerage in Las

Vegas, Nevada, where he specialized in Commercial Real Estate Investment and Tax Deferred Exchanging (IRS Code, Section 1031).

As often happens, Robert's marriage failed and he divorced from his first wife. After a brief relationship, which also failed, Robert met and married VaLoy Mathews, in 1975, in Las Vegas. Val, who she prefers to be called, was a native of Southern Utah, and the two began spending their idle time in that area, at her Father's ranch. Often, Robert's children, Danny (15), and Jeanne (12), accompanied Robert to the ranch. They came to love the area and often referred to Val's parents as Grandma and Grandpa Mathews. The two older children, Robert, Jr., (21) and Janis (20), were in school, at Las Vegas and Reno, and did not make it to the ranch. However, they had heard of it and often indicated that they wanted to go there. At this writing, Robert's children are quite successful today, with Robert, Jr. working at a large bank, in Delaware, Jan is, working on a Doctorate in Nursing, while teaching at the University in las Vegas, and Danny as the Project Engineer at the largest conglomerate of Resort Casinos in Las Vegas. Jeanne chose to marry and now is the mother of two fine children, while living in Las Vegas. All the others, except Janis, were, or are married, and have given Robert 6 Grandchildren.

Val had a daughter by a previous marriage, also. Donna (39, at this writing), has completed college, is now married and living in the Mobile, Alabama area, where she survived Hurricane Katrina. She has two children.

In 1992, Val's Father died, and, realizing that her mother, who had Macular Deterioration, needed help, Robert sold his brokerage, and moved to Southern Utah to help out. He obtained a Real Estate License in Utah and worked in home sales, briefly, but home selling in the Southern Utah countryside was nowhere near as interesting as Commercial Real Estate selling in Las Vegas, and Robert retired a second time, in 1987.

In 1988, Robert started looking for work, in sales, or Public Relations, where he discovered that his niche lied, trying his hand at several types of selling, including a stint as an owner of Barter Company. None of these were satisfying to him (there was too much underhandedness in the work) but he eventually took over as a route salesman for a First Aid Company, based in Missoula, Montana. Robert started selling with only 8 accounts in an area with scarcely 10 people per square mile, and, over the next five years, built it up to over 600 accounts, and, in 2002, was awarded third place in the nation by the Manufacturer of the first aid products and he and Val received a cruise as part of his reward.

In 1996, Robert retired for the third, and final time, due to disabilities, which he contracted during his military service and which, plagued him throughout the remainder of his life. Not satisfied with idleness, however, Robert became involved in Veterans' Benefits, and Veterans' Affairs. After reactivating his membership in the American legion, (he first joined the legion in 1961), and the Disabled American Veterans (DAV), he served in Various positions and continues to serve as the Adjutant of the DAV, Dixie, Chapter 7, in St. George, the 1st Jr. Commander of the DAV, Utah Department, in Salt lake City, and the Public Relations Person for the American legion, Lester Keates Post 90, in St. George.

At the turn of the 21st Century, Robert, who, up till then never told anyone of his military service at the age of 16, for fear that he might suffer from such revelations, heard about an organization known as the Veterans of Underage Military Service ( VUMS). He inquired as to their legitimacy, and received word that he had received a Presidential Pardon, and was declared a "Patriot", for having wanted to serve his country at such an early age. Robert joined that Organization and, today, serves as the Utah Commander, where he attends meeting across the Country, in an effort to represent the plight of Veterans whenever he receives the opportunity.

Several years ago, Robert, while writing a weekly article for the local daily newspaper, learned that there was a group formed, in order to develop a Memorial to the Veterans of Southern Utah at the St. George Cemetery, and upon completing that as well as sufficient monies to maintain the Memorial for the foreseeable future, and another Memorial, located at the Elks Club, in St. George, disbanded. Realizing that many more projects needed to be completed throughout Southern Utah to recognize the freedoms for which we, as Veterans, have fought, set out to reactivate the group. It became known as the Coalition of Southern Utah, and thrives today, to bring about such projects as a Nursing Home, a Veterans' Center, and, perhaps, a Veteran's Hospital, and Cemetery somewhere in the area. The group thrives today and is attended regularly by representatives from most of the Veteran's Groups in Southern Utah, as well as Representatives from the Veteran's Administration, the Utah Job Service, and State and Federal legislators.

Today, Robert continues to live in Southern Utah (not to be compared to Northern, (BRR) Utah, and continues to say, in deference to those who might feel otherwise, "To heck with Florida; Southern Utah is the 'HOLY LAND'!"

# James C. Hardwick, LM2538

*Age 16 - Texas National Guard*

Shortly after my 16th birthday in 1939, and while in high school in Dallas, Texas, I enlisted in the Texas National Guard, Troop A, 112th Cavalry. In the summer of 1940 we participated in a two-week maneuver in Louisiana at a location now known as Fort Polk. In 1940 it was a mosquito-infested wilderness.

In November 1940, the 112th was mobilized to active duty and ordered to Fort Bliss in El Paso, Texas. My age was discovered and I was honorably discharged before this mobilization.

An older brother was mobilized at this time and went to Fort Bliss with the cavalry. I then learned that one could enlist in the Navy at age 17 with parental consent. I joined the my 17th birthday, 5 December, 1940. The next few weeks were spent in boot camp in San Diego, California. The highlight of my San Diego experience was dancing with Rita Hayworth when she was a young starlet (I was in a conga line with about 50 sailors behind Rita).

Upon completion of boot camp I was given 7-days boot leave. I would have liked to go home, but on my pay of $21 a month, I could not afford the travel cost, so I passed up the leave.

I had applied for machinists' service school after boot camp and was pleased to learn that I would attend a new Navy school at the Ford River Rouge Plant in Dearborn, Michigan. Sailors attending this school lived in barracks within the Ford compound. In the morning we attended classes, mostly mathematics, and in the afternoon each sailor was assigned to a senior tool-and-die maker in the plant shop.

The UAW-CIO union struck the Ford plant during this period. Ford was the last of the automakers to be unionized and it happened during my time at the plant. I had the honor of meeting Henry Ford, Sr. while there.

Upon completion of school we were assigned to various ships in the fleet. I was happy to learn that my ship would be the light cruiser, USS Honolulu, then operating out of Pearl Harbor, Hawaii. I confess I did not know where Pearl Harbor was, but I knew it was far away from the war in Europe and the U-boat activity in the North Atlantic.

After a hectic train ride from Dearborn to San Diego, I then experienced a hectic sea voyage on a Navy transport ship from San Diego to Pearl Harbor. I believe there were 5,000 sailors on that transport.

My misery ended in Pearl Harbor when several of us were called out to be delivered to a beautiful light cruiser, USS Honolulu (CL-48), sitting at anchor in the harbor. The date was 6 June 1941.

My assignment on the Honolulu was to A Division in the engineering department. Responsibilities of A Division personnel were in the machine shop, evaporators, diesel engines, refrigeration and various other auxiliary equipment. My principal duty for the next five years was in the evaporators making the ship's fresh water. I also had the opportunity to stand watches in the main engine rooms and elsewhere to gain knowledge of most of the ship's mechanical systems.

From the time that I boarded the Honolulu until 7 December 1941, the routine was to leave Pearl on Monday to go to sea for war exercises and to return on Friday for weekend liberty in Honolulu. This practice was obviously well known to the Japanese planning a Sunday morning attack.

We had one pleasant trip stateside in September 1941, before the war began. This was to Long Beach, California. I received a 5-day leave which I spent hitchhiking to Bryce Canyon. Utah, Boulder (now Hoover) Dam, and the Grand Canyon. Very nice. Upon return to Hawaii, we resumed the old routine - to sea on Monday and back to Pearl on Friday.

The Navy had an R&R camp at Nanakuli Beach which, if you applied for a weekend there, your number might be drawn. My number was drawn for the weekend of 6 and 7 December. My 18th birthday was 5 December and I was completing my first year in the Navy. The night of 6 December we attended a luau at Nanakuli. Sleeping facilities consisted of squad tents for the attending sailors.

Early Sunday morning I awoke to the sound of many explosions coupled with the shout of a bosun's mate running down the tent row shouting for us to return to our ships at once, anyway that we could get there. We had no idea what was happening but a buddy and I followed orders and hitched a ride with a musician that had entertained us the previous night. He let us out on a promontory overlooking Pearl Harbor. The initial torpedo attack had occurred and the harbor was enveloped in dense black smoke.

We made our way into the navy yard unchallenged and went toward the Honolulu which was tied to a dock with the USS St. Louis. Some time during our progress through the navy yard a bomber scored a near miss on the Honolulu. The bomb exploded between the ship's hull and the dock, rupturing the hull and flooding a 5-inch-gun handling room. A short time later I boarded the Honolulu and went to my duty station.

The St. Louis, which was outboard of us, was the only ship that managed to get underway to sea that morning. Admiral Kimmel was aboard. Fortunately, they did not find the enemy.

We were suddenly transformed into a "dungaree" navy, which is to say there were rare occasions when we would wear a dress uniform for the next few years.

After repair of our damage we were assigned to go to the West Coast to convoy troops to Melbourne, Australia, and to Noumea, New Caledonia. In February 1942, we arrived in Melbourne. This was the first crossing of the equator for almost all hands on the Honolulu. The trip was uneventful, except for meeting a refuel tanker in the beautiful island of Bora Bora, Societe Islands. This was a very natural, primitive place in 1942. I understand it is now an overdeveloped tourist attraction.

Our next significant assignment was in the Aleutian Islands in Alaska. In August 1942, the Honolulu was in a task force assigned to bombard Japanese garrisons on Kiska and Adak Islands, way out on the tip of the Aleutians. Later, when the Army landed on Kiska, they found the enemy had evacuated. However, strong resistance was found on Adak.

After a visit to Mare Island Navy Yard in the fall of 1942, we were off again to the South Pacific. Our first stop was at Noumea, New Caledonia, where we saw what was left of the USS San Francisco after the mid-November Battle of Savo Sound off Guadalcanal.

The Honolulu then joined a task force consisting of other cruisers: the New Orleans, the Minneapolis, the Northampton, the Pensacola, and several destroyers. This force encountered the "Tokyo Express" resupplying their forces on Guadalcanal about midnight on 30 November. In the ensuing fight every American cruiser was either sunk or seriously damaged except the USS Honolulu. It was a moonless night and I believe it was one of the first engagements fought by radar. The first radar units were installed on the Honolulu in 1942. The engagement was called the Battle of Tassafaronga.

My assignment at Tassafaronga was in a damage-control crew stationed on the main deck under number 4 gun turret. When the battle commenced, a black night turned to daylight from the explosions, and the torpedoes were so thick I thought I could hear their motors.

Tassafaronga was followed by other principal engagements: the first and second Battles of Kula Gulf in the New Georgia Islands in 1943, and the Battle of Leyte Gulf in 1944. In the second Battle of Kula Gulf the Honolulu was struck by two torpedoes. About 70 feet of our bow was blown away by the first torpedo. The second one was a dud that only punched a big hole in our fantail.

On 20 October 1944, the Honolulu was part of the landing force at Leyte in the Philippines. The Army landed at about 1000 hours that morning and the Honolulu was on call-fire support for them. At about 1600 (4:00 p.m.) A lone torpedo bomber flew in low among the ships and launched a torpedo which struck the Honolulu 's forward port side, killing 62 men and seriously wounding many others.

Leaving the battle area severely damaged, we made our way to Manus in the Admiralty Islands where there was a floating dry dock. Some of our dead that could be recovered were buried at sea en route to Manus; the rest were recovered in dry dock and buried ashore at Manus.

This action effectively ended the war for us as the damage was so severe that we were ordered to the Norfolk Navy yard - almost halfway around the world. West Coast yards were very busy repairing ships less severely damaged that could be quickly placed back in action.

Repairs to the Honolulu took about 9 months. The atomic bombs were dropped and the war ended. After some visits to Newport, Rhode Island, New York City, Guantanamo, Cuba and New Orleans, the ship was ordered to Philadelphia to join the mothball fleet. The ship is credited with sinking five enemy ships, downing four aircraft, and supporting eight island landings during three years in the war zone. The USS Honolulu was scrapped in 1959.

I left the Honolulu in June 1946, almost five years to the day I first boarded her. The balance of my enlistment was spent in mothballing various ships in Philadelphia. I was placed on terminal leave in January 1947 and honorably discharged on 10 March 1947, after about 6 years and 3 months service.

Upon my return to civilian life I entered engineering school at Southern Methodist University in Dallas, Texas, graduating with a bachelor of science in mechanical engineering (BSME) in 1951. 1 became a licensed professional engineer in 1956.

I married Elva Pegel in 1947. Elva died on 25 September 2007. We have 5 children, 12 grandchildren, and 1 great-grandchild.

James Hardwick lives in Dallas, Texas.

# Barnie Henderson

*Age 15 - U.S. Navy*

I was born on July 20, 1927 just as the Great Depression was beginning. My dad worked in the Texas oilfields and we moved about the state often. When World War II began, I became a very patriotic American. About two years after the start of WWII, we were living in Houston Texas at the time when the original Cruiser "Houston" was sunk in February, 1942, in a naval engagement in the South Pacific. The city fathers of Houston organized a War Bond drive to raise enough money to buy a new Cruiser Houston, and the United States Navy had joined forces with the city to enlist enough sailors to man a ship the size of the new cruiser. A classmate and I cut classes on the original day of the bond drive and enlisted ... this got us a ride on a flatbed truck in the parade that was a part of the combined efforts of the city and navy. Later that day I was turned down for the first time for color blindness.

I was determined to become a member of the Navy and do my part in the war effort. Early in 1943, I made two more attempts to enlist, and on March 13th I made my final and successful try. Apparently my Navy Corpsman pulled me out of-the line and asked me if I hadn't attempted to enlist before. When I said "yes", he asked for my name and retrieved my previous enlistment records. He told me that if I was color blind it would not change and I was therefore ineligible to enlist. A Navy Doctor overheard our conversation and, because I insisted I could see colors, he told the corpsman to give me a "yarn" test, which I passed easily. I was sworn into the United States Naval Reserve on March 13, 1943. Although I was only 15 years old at the time, I was eager to do my part. I was sent to Farragut Naval Training Station for "boot camp" training.

Following graduation, several of our "boot" training squadron were sent to the Naval Air Technical Training station at Norman, Oklahoma to attend Aviation Ordinance School. I graduated in the top 10% of my class and was promoted to AOM3C.

After leaving Oklahoma, I was transferred to Whidbey Island Washington for aerial gunnery training, and then to Seattle. It took several weeks waiting to join the new Fighter Squadron 36 and begin the necessary crew training for potential carrier duty.

I spent most of the war doing the same thing — preparing air crews for ship board duty. Training consisted of several stages with the formation of a new squadron. Beginning in either Washington or Oregon, we would move down the west coast and end up at North Island, at San Diego. There, the squadron would board a carrier and head to sea — all but the nucleus of the next new squadron.

In 1945 I finally made it aboard the escort carrier U.S.S. Kula Gulf. This was a newly commissioned jeep carrier, which I boarded in Portsmouth Oregon and into Puget Sound for our brief "shake down" cruise. At this time, I was no longer in the fighter squadron, as I was assigned to the ship, as a part of the ship's company. We still performed the same tasks: cleaning and loading machine guns, assembling bombs and torpedoes, etc. Although I had trained earlier as an aerial gunner, I only made one flight at sea aboard a TBF. Landing on the deck of a jeep carrier was thrill to remember. I was unable to see where we were headed, as the gunner's position was in the ball turret facing aft! This was the same type aircraft that President George H. W. Bush was shot down in, early in the war. I was promoted to AOM2C while I Served aboard this ship.

The Kula Gulf never made it into combat. We sailed toward the South Pacific to become a part of the 5th Fleet. But we stopped off at Pearl Harbor on the way to take on supplies, and while there, Japan surrendered following the dropping of the second atom bomb. It was disappointing that we never really were "in the war." Although there were plenty sailors who would have traded places with us.

Our ship remained in the western part of the Pacific ocean for several more months, putting in at Ulithi Atoll, Guam and sailing into the typhoon that wrecked so many ships and island installations. We experienced some very heavy seas, lost the Captain's Gig which was lashed below the flight deck, which caused some minor damage to the hull of our ship as it was tom loose. We also had several aircraft which were tom from their moorings on the hanger deck, and wallowed around, causing heavy damage to the aircraft and the interior of the hangar deck.

In January 1946, I was transferred to Litchfield Park, Arizona, where I spent several months "pickling" brand new aircraft and storing them in the Arizona desert. When we accumulated enough points for discharge, we were sent to various separation locations around the country. I was sent to Camp Wallace, Texas; which was an old Army Base Being used by the Navy to discharge sailors. The most memorable thing about the three days I spent at Camp Wallace, was the final day when we were given our physicals. The medical personnel had an assembly line system that began by entirely disrobing, and going from station to station, to have our eyes, ears, heart, etc. checked out. When we got to the Proctologist's station, the Dr. had a sign on his desk that said "Super Duper Pooper Snooper". Imagine a room full of naked men laughing at that sign!

On April 6, 1946, I became a civilian again, following three years and one month of active duty and eight months of sea duty.

# Jerome "Jerry" Hendrickson

*US Navy September 1947 - 1956 AD 2*
*US Air Force September 1956 - 1967 SMSGT*

I was sixteen years old, working in the Joe Tittle meat and fruit market in Gary Indiana when a friend came in and said that he was going to sign up with the Navy and asked me to go with him. I hated Gary and though only sixteen years old, I had made several attempts to leave the town, and when he said that, It just struck me right as the proper thing to do to get out of Gary.

I convinced my mother that it was best for me at that time, and changed my birth certificate from the date of birth of 1931 to 1930.

We went to the recruiting office in Gary and got our date set to go to Chicago for physical and induction. There I was asked my age again, and responded that I was seventeen and nothing more was said.

I served eight years in the Navy, when one day, when I was assigned to VF 121, and aboard the USS Hancock, I asked our maintenance officer if he would find out for me, what I would have to do to set my age right in the Navy records. Two days later Ensign Ziener informed me that he thought that I should forget about it and say nothing more to anyone. The reason being is that the Commander was pressuring him to give up my identity so he could take action with court marshal proceedings.

In 1956 I took a discharge from the Navy at San Diego, and the following day went to Los Angeles California, where I enlisted in the US Air Force using my discharge papers from the Navy as proof of identity. So, at this point, I went in the Air Force with my records indicating my age one year older than the fact.

I was in the Air Force approximately two years when I confided in a personnel clerk that I was interested in knowing what regulations stipulated about a person changing his age from one that was used to gain entry in the service, to the correct age. He later informed me that he could find no repercussions resulting from my application to correct the records, as long as my records were clean, but I needed to prepare a written statement showing the cause of the error in the records, and it had to be submitted to Air Force Headquarters for approval.

So, I wrote up a statement that I had falsified my age to gain entry in the Navy and later used my Navy Discharge to join the Air Force. I Turned that in and the appropriate papers were submitted to the Air Force.

Not too long thereafter, orders were issued by the Department of the Air Force to change my records to reflect my actual date of birth.

I was proud of my services in the military and all the education and experiences that I gained, and it all lead to a good job with McDonnell Aircraft after retirement from the Air Force.

I since am retired after studying a few books on house construction, plumbing, and electrical, after which I purchased forty acres of land and physically built, plumbed and wired my retirement home with minimum outside help. My wife and I have enjoyed living here, outside of Lebanon Missouri ever since.

I devote most of my time to community service activities and have been a member of the Lions Club for over 20 years reaching the highest position of State Council Chairman.

## A BOY AND A STAR

As a mountain boy, "Buck" was forced to learn
How things can change, showing life, hard and stern,
When but a lad, he saw his dad breathe his last breath
His family circle was broken by the grim reality of death.

With his mom and his siblings, he crossed the ridges of blue
From Carolina to Tennessee in hopes of starting anew,
In Martin Chapel School, he read and dreamed and wished
By the Nolichucky River, he ran and played and fished.

While he was still in his teenage years
The armies of Germany were raising many fears,
The peoples of Europe were being devastated
As the madness of Hitler was being consummated.

Across America, thousands of young men
Were called into training, to bring the war to an end,
In his veins, his Scotch-Irish blood challenged his senses
He left his Unicoi home to join the allied defenses.

As the war raged on, he saw many friends die
In the foxholes he shivered as he watched the snow fly,
He longed for the days in the Tennessee hills
As the bombs were falling, making their kills.

Then came the day in June of 'forty four
The plans were made to invade France's shore,
The Normandy beachhead must be broken
On the Higgins boats, few words were spoken.

The German armies were wedged in tight
But these young soldiers were prepared to fight,
"Buck" and his comrades knew how great would be the cost
The dangers that lurked and the lives that would be lost.

They went in bravely in the face of fire
Hundreds died as the death toll climbed higher,
The Nazi resistance must end in defeat
These American Gl's would in no way retreat.

Buck fought bravely and well and traveled far
For his courage, he was awarded the Silver Star,
Thank God, he returned safely to his Tennessee hill
America can be proud of men like my uncle Will!

A Tribute to my uncle, William Buchanan

Jack D. Ham's

January 25,2001

# Joe C. Hernandez

*Age 16- U.S. Army Air Corp*

I was born on June 21, 1930 in San Antonio, Texas as the youngest of eleven children. It was just two years before the Great Depression and it was a tough time for my mother and father to raise a family. But just two years later, in 1932, my father passed away-leaving my mother with a houseful of young children to raise.

Around 1941, just before World War II, the burden on my family became even greater when one of my brothers drowned. This took a great emotional toll on my mother and my oldest brother, who were just trying to provide for the family. But soon after this, my five older brothers began getting draft notices in the mail to go join the army. I then had an overwhelming urge to join the army as well, despite the fact that I was too young, because I knew it was something I could do to help out.

In August of 1946, when I had just barely turned 16 years old, I decided to join the service too. When I went down to the office to join I had to lie and say that I was 18 years old. During the physical examination, I weighed in at only 98 pounds and was then told that I was not heavy enough to join the service. I had to plead for them to discretely add another 10 pounds onto my weight in the chart so that I could be considered eligible. Finally, I was admitted into the Army Air Corp at that time.

I was inducted at Fort Sam Houston Army Post in San Antonio, TX and was then transferred to Lackland Air Force Base for basic training. As soon as I finished up my basic training I received orders to Osaka, Japan with the U.S. Occupation Forces for 30 months. During that time I was asked if I wanted to go to cooking/baking school and, thinking that it would be a good opportunity for me to expand my skills, I quickly said yes. So I packed up and was off to the 5th quarter Master Cooking and Baking School in Yokahama, Japan. Once I finished my training I rejoined my unit in Osaka and by the time I returned to the U.S. I had been promoted to Staff Sergeant. Back in the states, I was assigned to Kelly AFB in San Antonio, TX and I reenlisted for another four years in August 1949. On March 9,

1950, when I was 19 years old, I married my high school sweetheart, Helen. But as the Korean War broke out I received orders to transfer to Norton AFB in San Bernardino, CA. I wasn't there for more than a year when I was shipped back to Japan for a tour of 30 months and Brady AFB in Hakata, Japan. I had to leave my new family and pack my bags to go help my country. At Brady, I continued to work in food service as NCOIC of the Snack Bar and after my 18th month there my wife (Helen) was able to join me in Japan.

At the end of my 30 months I went back the U.S. and reported to the 750th AC & W Squadron in Boron City. There, I was assigned as NCOIC of the dining hall. Shortly after, I returned to Norton AFB and remained there until October 1958. Then my supervisory work in food service took me to Mt. Laguna AFB in Mount Laguna California where I remained until March 1960. I was then back to Boron to rejoin the 750th AC & W Squadron where I was promoted to Tech. Sergeant. I remained in California until July 1963 after also serving at Santa Rosa Island. My time in California was done and I received orders to KI Sawyer AFB in Marquette, Michigan. I was only in Michigan for a year when I was transferred back to Lackland AFB in my hometown of San Antonio, TX and I remained there until the end of my military career in September 1966.

I am grateful for the many opportunities my service has given me. Joining the military made it possible for me to provide for my family, and I am fortunate to have been able to serve my country in return. I have not been back to Japan since my time spent serving there. I am still married to my wife Helen. We had three children and we have now been married for 59 years. We have four grandchildren and our first great-grandchild on the way.

# John W. Hill, LM2160

*Age 15 - United States Navy*

I was born in Shamrock, Texas, on 18 July 1928. My dad was a sharecropper. We lived in Shamrock until I was 3 years old, then we moved to Lubbock, Texas. This was right in the middle of the Depression. I sold newspapers and shined shoes to help Dad support the family.

There were five boys and four girls in our family. We hoed and picked cotton just to get something to eat. On many occasions we went to school with nothing to eat, and sometimes even without shoes on our feet. I remember walking to school barefoot during a light snow storm.

It was a hard life and my dad began to drink, and as time went by he became a drunk. At the age of 14, I dropped out of school in the eighth grade and took a man's job in order to support the family. I finally got tired of home and my dad's ways, so at age 14, I attempted to join the Marines. I was rejected because I had a police record for a minor juvenile indiscretion.

When I turned 15, home life became more unbearable. A friend told me how I could get into the Navy. I falsified my birth certificate to hide my age and at 15, I joined the Navy.

I was sworn in on 7 February 1944 and left Lubbock for San Diego for boot camp. After boot camp, I was sent to Treasure Island, California, and in April 1944 we left for the South Pacific on a merchant ship, the SS Santa Monica. We sailed for Noumea in New Caledonia, near Australia, where we went aboard the hospital ship USS Tyran.

I was a seaman and a second powderman on a 5-inch gun. We left for Guadalcanal, Bougainville, and Talasea in the South Pacific. After Bougainville, we sailed on to Auckland, New Zealand.

After serving 3 months on the USS Tyron, they found out about my true age and I was put off the ship in Noumea and sent to San

Francisco for discharge. I was sent home to Lubbock, Texas.

After I returned home I met a girl with whom I had gone to school, and three months later, we were married. I was 16 and she was 15. Some people said that the marriage wouldn't last, and that we would still not be married six weeks later, but that was 62 years ago last November. We have 5 children, 8 grandchildren, and 5 greatgrandchildren. Some asked why I married so young, and I said I was afraid someone else would get that pretty girl.

I went back to school, finished my GED, and went on to college. I was called to preach in 1956. I am now a retired minister, but we spent over 50 years preaching and pastoring churches. I am an ordained bishop in the Church of God. I was a member of the Navy reserves for 18 years and am a life member of the VUMS. Life has been good and I am very blessed.

John Hill lives in Saginaw, Texas.

# Fred L. Hoke, M2619

*Age 15 - United States Army Reserve*

I was born in Thermopolis, Wyoming, on 21 May 1940. When I was young my parents moved to Mountain View, Wyoming, about 20 miles from Casper, Wyoming, where I lived until entering the Navy.

I was the youngest of four children. My parents were very poor. Our house was made of railroad ties. We did have water piped into the house. To get warm water, it had to be heated on the stove. Our toilet was an outhouse. We had a radio, but no television. Each fall I would get a new pair of shoes (large enough to grow into) and a couple sets of clothing for the new school year. I learned early in life to take care of what I had and not to lose anything. My father was a trained blacksmith and was skilled at shoeing horses. When the automobile replaced horses his skill was no longer needed. He worked as a laborer on any job he could get. My mother was an excellent cook, her full-time job, besides cooking at home.

My folks were excellent farmers. They had a large vegetable garden and my mother canned vegetables. We raised chickens, which kept us in eggs and meat. We ate a lot of venison and antelope. I'm sure many game wardens turned their heads because they knew it was a necessity for us. I grew up with miles of open land to roam and always had a rifle in my hand. I never killed anything that we wouldn't eat. I once watched a herd of about 1,000 antelope cross over a hill. It was an amazing sight to see!

As I was growing up I did a little of everything around our place, helping as much as I could. When I was 13 I won a gold pen set, a ride in an airplane, and a new bicycle, for being an excellent newspaper boy. My route started out with 30 homes and grew to 300. When I completed my ninth year of school I worked part time at a grocery store. I was offered a full-time job and I took it. I was earning more than my dad at that time. I had a second job at a bowling alley.

When I was 15 I talked my mother into backdating my age, and I joined the 349th Armored Field Artillery of the U. S. Army Reserve. This was on 29 November 1955. At that time I planned my military career. I would stay in the Army for a time, then join the Navy. I attended meetings one weekend a month and spent a month at Camp Guernsey, Wyoming, for training. Sometime in 1956, my unit was called up to help fight forest fires that were threatening homes in Cheyenne and in Nebraska. For a month and a half, I drove a truck pulling a 2,000-gallon tank-trailer. On one trip I was the last one in the convoy. Toward the top of a mountain, driving over very rough roads, something came loose and I lost the 2,000-gallon tank.

I received my Army discharge and on 26 May 1957 and I was sworn into the Navy. My goal was to obtain the rank of commissioned warrant officer before my 38th year. I was sent to boot camp in Great Lakes, Illinois. After boot camp, I reported on board the USS Los Angeles (CA-135). I was assigned to M-Division in engineering.

In late 1958, the Los Angeles was part of the task force that stopped the Chinese from invading the islands of Quemoy and Matsu. In 1960, the Los Angeles went up the Saigon River in Vietnam for the first surge of war in Vietnam. During the Cold War we were about 800 miles from the coast of Japan, lying dead in the water. It was a standoff with a Russian submarine which had been following us. We went on high combat alert and made ready our missiles. The standoff lasted 14 hours. There was no communication with the Russian sub. Finally, our captain sent a message giving them five minutes to communicate with us or we would blow them out of the water. Just before the time was up we heard the propellers of the submarine and it disappeared, to our great relief. We would probably have gotten them but they would have had at least one shot at us.

I was discharged from the Navy, but reenlisted in less than a year. This time I was sent to the USS Watchman (AGR-16). We were a spy ship; our patrols were out about 1,000 miles from the East Coast. Our ship was loaded with radar, sonar and other listening devices. This was the time during the Cold War when Russian ships were everywhere. Our patrols lasted 30 to 45 days, or until relieved by another ship. I spent four years on the Watchman. By this time the Vietnam war was going strong.

I enlisted again and was sent to the USS Ranger (CVA-61) in port at the Alameda Naval Base, California. I was assigned to the M-Division and we were soon on our way to Vietnam via Pearl Harbor. On 6 August 1964 we arrived at the Gulf of Tonkin, Vietnam, at the time the USS Maddox and the USS Turner Joy were being attacked by North Vietnamese motor torpedo boats. The USS Ranger struck back at them, hitting naval bases and other land targets. I made three trips into Vietnam's combat area. During the last tour we lost 29 comrades and our engine room caught on fire, losing a man there. No one wanted to work the engine room. They felt it was a bad omen.

At times I would go out on patrol boats to help diagnose mechanical problems. We clashed with North Vietnamese torpedo boats several times. We got shot at, won some battles and lost some. My longest tour in Vietnamese waters was 16 months. During this time we had many planes shot up and many downed. I have lots of sad memories.

My goal that I set when I first went into the Navy finally came through. I was to be promoted to warrant officer. It was sad, but I had to turn it down because of serious family problems. I received an honorable discharge as a machinist mate first class (E-6) after serving a combined total of more than 11 years. It took me many years to pull myself together after Vietnam.

After my discharge I worked for the Navy for one year and the Air Force for more than 14 years. I received a number of awards as a civilian worker for the Air Force. I am a proud American and proud of my service to my country.

Fred Hoke and his wife Mary live in Apache Junction, Arizona.

# Wendell J. Johnson, LM2428

*Age 15 - United States Navy*

I was born on a farm in the small town of Hankinson, North Dakota, on 8 August 1927. I was the youngest of three sons born to Hannah and Martin Johnson. My father, who came from Sweden, joined the U. S. Army during World War I. When he became ill and could no longer farm, we moved to town. This was during the Great Depression. All we had was my father's army pension, a small garden, and a cow. During the winter of 1936 the weather was so cold that my parents decided to move to California. We settled in Menlo Park, California. My father could no longer work, so my mother held down various jobs including working at the Stanford University library. I delivered newspapers, cleaned horse stalls, and did gardening work.

When World War II began, my brother and I, like many young men, were anxious to join the service. He was accepted into the Army at 18 and became a paratrooper with the 82nd Airborne Division. I was anxious to become a paratrooper like my brother, so I erased the birth date on my baptismal certificate and made it 1925 instead of 1927. I went to the recruiting office in Redwood City, California. Recruiters from the various branches of the service sat around a large table. I first went to the Marine recruiter because I knew they took men at 17 and had paratroopers. He told me that their quota was full at the time, so I next turned to the Army recruiter as I still thought I could be a paratrooper, but he told me they took men at 18 only. The Navy recruiter had heard me asking about the paratroops, so he said that they had parachute riggers and they would take me at 17, so I signed up with the Navy. Needless to say, I never became a parachute rigger.

I was sent to Farragut, Idaho, for boot camp and from there I went to San Francisco, California, where I was assigned to the destroyer USS Stephen Potter (DD-538) on 21 October 1943, the day the ship was commissioned. After a shakedown cruise, our ship was assigned to Task Force 38 and later to Task Force 58.

I remained on the Potter until the end of the war. My main job was as part of the deck crew. When we were at our battle stations, my job was gunner and loader on the 20-mm antiaircraft gun. I remember writing my mother' s name, Hannah, on the steel plate of the gun.

Around Easter Sunday of 1945, I saw the aircraft carrier USS Franklin hit by enemy bombs. She burned for most of the day and night. There were 700 men left out of a crew of 3500. I shall never forget the sight and sounds as our ship pulled alongside. There were a few men with musical instruments playing the song "Easter Parade," to help keep up morale. Whenever I hear that song, my memories go back to that sad day.

I watched as the aircraft carrier USS Bunker Hill was hit. We picked up 107 of her crew that day in May 1945. Many men, full of shrapnel and bullet wounds, were jumping into the water. Many were badly burned with skin hanging from their faces and arms. I saw 300 men who were killed, their bodies wrapped in the hammocks and lined up in a row that stretched from one end of the flight deck to the other. I watched as men slipped a projectile into the hammocks of the dead to help them sink to the ocean floor when they were buried at sea. These sights remain in my mind to this day.

We supported the troop landings on Iwo Jima.

I saw many men die there. We were under constant air attacks from torpedo planes, dive bombers and kamikazes. We had a close call while I was at my battle station. A bomb was heading straight for us. We ran for cover as the bomb landed on the other side of the ship.

During one of the worst typhoons in the history of the Pacific, between the island of Ulithi and the Philippines, three destroyers, the Hull, the Spence, and the Monaghan capsized. About 790 men were lost. Our ship recorded a number of 70 degree rolls. Many of our men were in life jackets expecting to go into the ocean. Even a little dog we had on board had on a life jacket. I was not on duty during the worst part of the storm. I was below deck praying that God would spare us, and He did. Our ship earned 10 battle stars during World War II.

At the end of the war I was sent to Treasure Island, California, for reassignment. I was then sent to Subic Bay in the Philippines and assigned to the sub-chaser 723. I was finally discharged at Camp Shoemaker, California, in March 1946.

I went back to high school and graduated in June 1947. I attended San Mateo Junior College, graduating two years later. While in college, I joined the National Guard.

I joined the U. S. Army in 1949 and finally became a paratrooper. I received basic training at Fort Ord, California, and then was sent to

Fort Benning, Georgia, for jump school. I also received training in gliders. After graduation I was sent to Camp Campbell, Kentucky, where I was assigned to the 511th Infantry Regiment of the 11th Airborne. Division. I made 7 jumps, 5 in training and 2 with the 11th Airborne Division. I was discharged in March 1950. Some months later, after the Korean War broke out, I sailed with the Military Sea Transportation Service on the General E. T Collins. We took troops to Japan and to Pusan, Korea.

It was during that time that I met my future wife Carolyn. We were married on 9 September 1951. I went to work for the Peninsula Building materials in Menlo Park, California, that year and remained with that company until retirement.

As of this writing God has blessed us with 57 years of marriage. We have 1 son, 2 daughters, 10 grandchildren, and 16 great-grandchildren. Our son was also a paratrooper with the 82nd Airborne Division.

I was never proud of my dishonesty in getting into the service, but I have always been proud of serving my country and helping to preserve the freedoms we all enjoy.

Wendell Johnson lives in Vancouver, Washington.

# Raymond C. Kaiser, LM2453

*Age 15 - United States Army*

I was born in Louisville, Kentucky, on 24 July 1932. When I was 15, school was out and I was hanging out with friends. We were all bored. One of the older guys said, "Let's join the Army." It was not because we were patriotic, it was just something we could do. There was no reason for us to enlist, and no one gave it a second thought. We were just dumb kids with nothing to do.

I told my friends that I wasn't old enough; I was just 15 and the youngest of the three. The older one said, "Tell them you are 18," and I said, "Okay." We went to the recruiting office and told them we would like to join.

The first question the recruiter asked us was, "How old are you?" I told him that I was 18. He asked for a birth certificate and I said that I didn't have one, and that I was only 17, not 18. He told me to get a notarized statement from my parents as to my age and I could then enlist.

I went home and begged my mother to sign and let me enlist. She wanted me to go to school and become something. My older brother had joined when he was underage. I wanted to do the same. She gave in and I returned to the recruiting office.

I was tested and told that I had high scores and was asked if I would be interested in going to school to learn high-speed Morse Code. I said, "Okay," but I had no idea what he was talking about. He told me that once trained, I would be assigned to a special organization. I still didn't know what he was talking about.

They said that I weighed 130 pounds, but I weighed less than that. I think they bent the rules to make a quota. I was 5 feet 7 inches, but by the time I was in the Army six months, I was 5 feet 11 inches and needed a new issue of uniforms.

I was sworn into the Army on 29 June 1948 and was sent to Fort Knox, Kentucky, for basic training. The next morning we were hustled into a formation to be inspected. When the sergeant came to me he asked if I had shaved. I said that I had. He said for me to go shave

NOW! I started to go and stopped. I said I didn't have a razor. They found a razor for me and I shaved for the first time in my life. They should have known that I was underage.

After basic I was sent to Fort Monmouth, New Jersey, for high-speed Morse code training. First, I had to learn to type. I told the sergeant that I couldn't type. He said that I would learn by the end of the week or else! I got the idea and was pretty good by the Raymond Kaiser end of the course. I never did master Morse code; I didn't want to. Later in my career I found these guys who had to copy code 8 hours a day, 5 days a week, to be a little nutty.

My friends and neighbors thought I had gotten into trouble because strangers were asking questions about me. A security background check was run on me because of the unit I was being assigned to. I found this out when I took Christmas leave.

I was sent to Germany on 5 February 1949.

Since I didn't finish Morsecode training, I was given on-the-job training into another military occupational specialty (MOS 805).

Some of my friends knew that I was underage, and I think they looked after me. When I turned 17, I thought it was okay to have my age corrected on my records. I was told I could stay in or take a minority discharged. I liked the idea of going home, so I got my discharge. I stayed out for about 3 months, then re-enlisted on 10 January 1950, and retired on 1 August 1970, after 22 years of service. I was in Korea during the Korean War. I did two tours in Korea, Japan, Germany, and Panama.

I said that I did not go into the Army for patriotic reasons, but I'm very patriotic. I would do it all again if I could.

Raymond Kaiser lives in Louisville, Kentucky.

# Michael G. Kapnas
# "Seaman Recruit to Commander"

*Age 16 - U.S. Navy*
*How a caring teacher and World War II*
*changed a young boy's life*

Michael G. Kapnas was perhaps the youngest teenager to reach Naval officer status during World War II. Through a strange turn of events and hard work, he became an Ensign, or entry level officer, in his 18th year.

Kapnas vividly recalled the day he heard the announcement that Pearl Harbor was attacked. He was just 14 and working as a movie usher in his hometown of Gary, Indiana, which borders Lake Michigan. In the middle of a screening, the film stopped, the lights burst on and an announcer's voice came through the theater: "The Japanese have just bombed Pearl Harbor."

"Where is Pearl Harbor?" Kapnas heard theater goers murmur to each other. As the news sunk in over the days that followed, Kapnas realized he would never forget where Pearl Harbor was, despite the thousands of miles and great ocean that separated it from Indiana. Little did Kapnas know he was destined to visit Pearl Harbor in the years to come.

Kapnas wanted to help his country, but knew he was much too young to enter the military. Recruits had to be at least 17 and, at that adolescent age, would also need consent from a parent.

A year and a half later, as a 16-year-old sophomore, the war continued to rage and Kapnas learned that a national test to join the armed forces would be proctored at his high school. He wanted to try even though, as a sophomore, he thought he didn't have a chance and wasn't considered eligible.

Fortunately, a kind high school guidance counselor, Elsie Cooper, would be the proctor. Ms. Cooper, who Kapnas remembered as a caring and supportive teacher, encouraged him. She said she would set up a test for him with the large group of juniors and seniors "just to practice."

"Next year, when you're eligible, maybe the exercise will help you pass," Kapnas recalled Cooper saying.

The test was tough and only five students triumphed, including, to his surprise, Kapnas. The result proved quite a predicament: Kapnas received orders to take a physical at the Navy Recruiting Center in Indianapolis,

Indiana, some 150 miles from his hometown.

Should he go? Was he eligible?

He felt a sense of duty and obliged the orders. Kapnas passed the physical and was astonished to be immediately sworn into the Navy on April 23, 1943, as a Seaman Recruit. No one mentioned his age, so he didn't either, he remembered.

But he did explain to the recruiting officer that he had already enrolled in summer classes and needed the credits if he were to become a high school junior.

It was agreed he would finish summer school and report to active duty on November 1, 1943.

But there was another obstacle: as a minor, at least one parent had to sign a consent form and his mother was strongly opposed. His father, however, had served in the Greek Navy during World War I and was happy to grant permission.

Yet less than two months later, Kapnas received orders to report to Northwestern University in Evanston, Illinois, on July 1, 1943, one week after his 17th birthday.

At Northwestern he entered the first class of the World War II U.S. Navy V-12 Program, which was designed to provide young men whose families had endured the Great Depression with a college degree in return for serving their country. When the program ended in 1945, more than 60,000 of the program's officers had served in the Navy and Marine Corps, according to the book "Navy V-12" published by Turner Publishing Co. An estimated 60 percent of the nation's youth in the draft age category took the qualifying exam, the book reports.

In the fall of 1944, a little over a year after he entered Northwestern. Kapnas had completed college credits that equaled almost three years of study.

He foresaw graduating soon and finally joining the fight abroad, when his Captain called him into his office . "It seems you enlisted under false pretenses," he said, and gave Kapnas two choices: he could be discharged as an "underage fraudulent enlistment, or he could agree to change the date of his enlistment from April 23, 1943 to two months later on the date of his 17th birthday, June 24, 1943. For Kapnas, the choice was clear; He changed the date.

Then, on November 1, 1944, he was sent on to midshipman training at Abbott Hall in Chicago. Finally, four months later, on March 29, 1945, just months before the war ended, Kapnas was commissioned to the rank of Ensign at the age of 18 years and 9 months.

Then it was off to General Line School in Miami, Florida, and amphibious training in Little Creek, Virginia.

Kapnas never fought in the war, but served the Navy almost 30 years.

In the two years after he became an Ensign, he skippered three ships) LCT 602, and LCT 538 and a yard freighter, YF 864, on home soil and trained crews in amphibious, or water-to-land, procedures to prepare to invade Japan. He was stationed in Virginia and Green Cove Springs, Florida.

Usually, an LCT would have two officers, but because of the Second World War, the Navy was short of officers and the teenaged Kapnas would be the only officer and skipper aboard his ships.

He recalled that one of them, the LCT602 he skippered from the Brooklyn Navy Yard to Norfolk, Virginia, had just arrived from the Battle of Normandy fresh with war scars - huge iron plates welded over shell holes.

But' there was another ship Kapnas will always remember; it was the yard freighter, YF 864, he commanded from Florida to Casco Bay, Maine. As the ship came upon the New York Harbor, the fog became so thick that Kapnas ordered the crew to drop anchor. They had no radar to guide them in the haze, and weren't exactly sure here they were. Through the night, as he stood on the bridge with his Quartermaster manning the fog horn, Kapnas could hear the "whish, whish" of the huge tankers that seemed to barely miss his much smaller ship. As daylight broke through the fog, he looked up to see the face of the Statue of Liberty towering ahead in the distance, and had to laugh - and breathe a sigh of relief- at how close they were to the Brooklyn Navy Yard, where they had planned o stop and refuel, he said.

In that period, also, Kapnas was awarded the Navy Commendation Medal for assisting several Landing Craft Tanks from broaching on nearby shores during a hurricane.

In the summer of 1946, Kapnas was assigned as LCT Type Commander, 16th Fleet, which consisted of more than 60 LCTs. He was just 20. He took command of LCT538 as Commodore, and his landing craft became the designated flag ship of the fleet.

In November 1946, Kapnas was released from active duty. He returned to Northwestern University on the G.I. Bill in 1947 and received a bachelor's degree in August 1948. He married his lifelong partner, Helen I. Tsatsos, of Hammond, Indiana, on June 5, 1949. The couple had met at a dance in Gary, Indiana, in 1946. They became engaged in 1947, but their parents said they could not marry until Kapnas finished college.

Even as a civilian, Kapnas continued to serve the Navy in the U.S. Naval Reserve. He felt he owed the Navy for the opportunities it afforded.

He remembered checking the newspaper every day to see if he would be recalled to the Korean War and, later, Vietnam. The general mood was that it could happen at any time, he said. As a reservist Kapnas' would go on active-duty for training missions at would last between two and eight weeks, induding a six-week tour aboard the U.S.S. Missouri (BB-63).

At one point Kapnas sailed a destroyer from Long Beach, California, to Pearl Harbor, Hawaii, the site he had heard about years earlier.

In 1967, he received a special award from the Chief of Naval information for public relations achievement. Only three of those awards were given nationwide that year, he remembered.

Kapnas retired from the Naval Reserve in 1971 in the rank of Commander. His civilian career that followed included various management positions with two major steel companies, U.S. Steel and Inland Steel.

At the age of 47, after 25 years in the steel industry, he moved to a third company, Whiteco Industries, where he was named the director of public relations. The firm was one of the largest outdoor advertising companies in the nation that also owned a major hotel and entertainment center near Chicago.

In 1980, at age 54, Kapnas left Whiteco to form his own public relations firm, Michael G. Kapnas Associates Inc. He retired with his wife eight years later to La Costa, a charming community in Carlsbad, California, a few miles north of San Diego. Kapnas has assumed countless leadership roles throughout his life both his career and volunteer organizations. He credits the Navy teaching him to become a leader.

In 1964 he served as the president of the Bailly Junior High School PTA when white and black students were integrated without incident, he said.

He served as chairman of the Gary City Plan Commission and the board of Zoning Appeals and president of the Merrillville Chamber of Commerce. He planned events for the March of Dimes, Muscular Dystrophy and the American Cancer Society, as well as citywide parades for Armed Forces Day and Memorial Day.

Kapnas said that his capacity for leadership was born and honed by the Navy as a young man. Serving his country taught him the value of contributing to his community. Even in Carlsbad as a retiree in 2002, he and his wife were invited to participate in the Carlsbad Citizen's Academy, a program designed to link residents with local government. Kapnas also served on his homeowner's board in Carlsbad for a number of years.

Kapnas grew up in a small town as a first generation child of immigrants from Greece. Without the V-12 program, he most likely would not have gone to college, he said. His future started with one caring teacher and some luck, and continued with a good ol' American work ethic.

# Peter M. Kniskern

*Age 13 - United States Army Air Forces*
*Age 16 - United States Marine Corps*

I was born in Hinsdale, Illinois, on 4 May 1929, just before the start of the Great Depression. My father was a veteran of World War I, and my grandfather served in the Civil War. My mother was French-born. Like many of us, I suffered through the after-shock of the Depression which included broken and separated families, divorce, work trauma, loss of jobs and income, and a nine-to-ten-year separation from my father, my country, and my childhood friends. This hiatus in my life placed me in Tunisia, the "Land of a Thousand Nights," with all of the exciting adventures that ensued.

In 1934, the year of the Chicago World's Fair, I moved with my mother and my sister Antoinette, from Hinsdale to Europe and then to the arid Sahel region of Tunisia, North Africa. We joined other members of the family on their plantation near Sidi-Bou-Zid, a major agricultural center. My two uncles were engineering agronomists who had been invited into the country by the Bey of Tunis to help curb the famine and help reclaim productivity in the Sahel. Many professionals came from France, Borneo, and Sumatra in the Pacific, all receiving land grants from the Bey.

I was enrolled in a French school at La Marsa, near the site of ancient Carthage. I not only learned to speak French, but to think in French. It was a cosmopolitan group of children that I associated with. There were French, Italian, and German boys.

During the time we lived in Tunisia, I became quite familiar with the country and its people. In addition to French, I could also manage quite well in Arabic. My linguistic skills and knowledge of the country would soon come in handy.

When the German army moved into Tunisia from the Lybian desert in 1942, our family had to escape. We went to Algiers, which was being bombed nightly by the Axis Air Force, and where life became quite hectic and interesting. The Allies invaded North Africa in November 1942 and soon occupied Algeria and the countries to the west. My knowledge of the region and its people, along with my fluency in French, caught the attention of U. S. Army officials. I was signed up as an interpreter and liaison for the 12th Army Air Force. As best as I can recall, this was in late November or early December 1942. I was 13 at the time, but my size and demeanor fooled most people most of the time. I was issued a U. S. Army uniform and was paid $30.00 per month. My direct employer was then Captain Stuart B. Mockford, USAAF.

In addition to serving as an interpreter, I was appointed as a civil affairs officer. My duties were to guide military officials in their search for proper billeting, and to assist in their daily objectives and other sundry assignments. Since I was a minor, they had to make it official and pay me a monthly salary.

In early 1943, while traveling to the interior of Algeria, I was mistaken for a German spy. After finally locating my immediate employer, Captain Mockford, I was released to his care and issued a pass to travel by military convoy to Constantine, a major central business and agricultural area in southern Algeria. This area was also occupied by American forces. In order not to be arrested again, I was instructed to wear a USA armband so I would not be stopped. However, this did not keep me completely safe. I was interrogated several times because other officials had not been informed about an American minor wearing the uniform and working for the U. S. Army Air Forces.

Another time, I was on my way to visit my family in Corneilles; Algeria with Tunisia as my ultimate destination. Many family members had fled to the mountains in the Corneilles region where one could feel relatively safe from the Germans. On the way there, as fate would have it, I got into trouble again because I had failed to put my armband on. I ran into some troops whom I had met on my initial trip to Algiers, and I accepted their offer to accompany me the next day to my intended destination.

We left the next morning. Along the way I noticed that the train was heading north, not south! I became increasingly worried and asked the officer in charge where we were headed. He finally admitted, "We are heading north, and you are being arrested as a German spy, so stand by!" Of course, I knew that this was a big farce, so I laughed. I should have remembered that by removing the armband, I had increased the chances of being arrested threefold. The situation became more tense, and I finally began to truly worry about my situation. This wasn't the way the trip was supposed to end; it was supposed to be a family reunion, not an occasion for my arrest.

The train arrived in Algiers in the afternoon. I became weak in the knees at the sight of four military policemen with Tommy guns at port arms who had been sent to the station to escort me back to headquarters. I said, "What a mess I'm in really, I'm not a German spy! Captain Mockford will corroborate this." So they tried to trace him though headquarters. He was finally located on temporary duty in Constantine, in the southern part of Algeria. When he was told the circumstances of my arrest, all I could hear at the end of the line was loud laughter and choking. Later I was released and joined Bill in Constantine. Bill Mockford was a good and great guy. He was one of the nicest officers I had met.

I served the interests of the Army Air Forces for several months, and there were times when my youthful exuberance got the better of me. I was visiting a pilot friend of mine when the base was attacked and bombed by the Germans. We heard the air-raid sirens, and most people ran to the shelters. But as a kid, naturally I ran to the top of the building to see the fireworks rather than run down and hide in the cellar. During the raid, one of the bombs did land and explode relatively close by, and some shrapnel did bounce off my helmet. When the commanding officer heard about my stunt, he ordered, "You are never to do that again!" I replied, "Yes, sir."

We went back to Tunisia after the Allied victory there, but like so many American soldiers, I longed to see "the old lady," meaning the Statue of Liberty. As I was a minor, the Army, with help from the State Department, fixed it up so I could return to the States. I can't recall exactly how the actual separation took place. I went by plane from Tunis to Algiers and on to Casablanca, then by ship to Staten Island, arriving there on 16 August 1943, and then on to join my father in Cleveland, Ohio.

I arrived back in the States to a hero's welcome and to a West Point appointment which I had obtained through my family connections. Regretfully, I arrived in the States too late to take advantage of this wonderful opportunity. I was interviewed by The New York Times, The Chicago Tribune, The Los Angeles Times, and others. However, when it came time to resume my studies, I discovered that I had matured beyond my peers. I wasn't the hero to them that I had hoped to be; except for the girls, of course.

Later, I attended the Tennessee Military Institute Prep School for two years before volunteering for the U. S. Marine Corps at age 16 1/2. I completed high school through the U. S. Armed Forces Institute during my years in the Marines. During my very instructive and informative stint in the Marine Corps, one of my most constructive assignments was when I assisted the Navy with some translations during the atomic tests at Bikini Atoll.

After my discharge from the Marines, I started my college studies at the University of Georgia, remaining there for two years. I then entered the *University de Grenoble* in Grenoble, France, and obtained my *Certificat d 'Etudes Commerciales en Droit et Langues.* I returned to the States and finished my education at George Washington University, majoring in linguistics (Spanish, French, German, and Italian), art, literature, international law and commercial trade, and economics.

I am currently the owner, president, and chief executive officer of Hemispheric Business Network, Inc., of Oakton, Virginia, an international network marketing company, which I purchased in 1989. Prior to acquiring Hemispheric Business network, I was the Washington liaison and representative for the Fils De A. Pouey, a company in Bordeaux, France, and was also completing my thirty years of association with Sun Financial Group, specializing in assisting the diplomatic and business communities.

At this time of my life I am interested in serving my country once again. My objective is to somehow assist the President with international relations. With my knowledge of languages and experience with foreign nationalities, I believe that I could contribute to a damage-control initiative with our Allies.

Peter Kniskern has lived in Oakton, Virginia, for the past 30 years.

# Roy Lee

*Age 15 - U.S. Navy*

During school summer breaks (1940 and 1941) to help make money for the family, Roy Lee (aged 12 and 13) worked aboard sea-going tugboats, shipping out of Galveston, Texas qualifying him as Merchant Marine.

When America was brought into World War II an uncle prompted the young lad, "Son, if you're gonna be at sea it may as well be aboard a ship that can fight back." So at age 14 Roy headed off to enlist in the Navy. Because he had a Copeland book (from service on the tugs) that listed an older age, he was immediately sworn into active service. A very short boot camp and off he went to war. Roy served mainly in the Pacific theater from the Aleutian Islands to the Philippines. He and a crew picked up the brand new ship PC1081 (Hell Harvest) from Albina Shipyards in Portland, Oregon. After receiving its commissioning pennant they set out to sea. On the 'shake down' cruise the shake down officer was actor Robert Montgomery, a hellava sailor.

Service in Nets and Mines was to be his longest duty. Laying submarine and torpedo nets and mines in strategic locations vital to the safety of American held harbors. And he was no stranger to dodging bullets, mortar rounds and submarines either. When the end of war was declared, Roy was preparing for MARS 1 - the invasion of the Japanese mainland to take place on November 1, 1945 (his 18th birthday). An estimated one million Allied casualties were expected in the first attack wave.

Three days after he turned old enough to enlist, Roy Lee was honorably discharged from active duty at Naval Air Station in New Orleans, Louisiana, November 4, 1945 with enough points for himself and two more to get out of the service. Roy remained in the Naval Reserve through the Korean War.

December 2005, Roy was appointed Louisiana State Commander of Veterans of Underage Military Service. He is a member of Veterans of Foreign Wars and the American Legion.

# Clifford Lowe, Jr., LM1122

*Age 16 - United States Coast Guard*

I was born in Charlotte, Michigan, on-14 April 1926. My parents divorced when I was about 3 years old. I lived with my mother. We lived in 12 different houses by the time I was 14 years old. My father lived in Toledo, Ohio.

I was washing dishes in a restaurant on 7 December 1941. We had the dinner dishes done and I sat at the counter reading the Sunday funny papers. A man came in and said, "The Japs just bombed Pearl Harbor!" I didn't know where Pearl Harbor was. Some of my buddies that were of age joined the military and I wanted to join too. I thought we could beat the Japs in a couple of years and come home.

I was out of school on summer vacation in June of 1942 and was ready to join, but Iwas only 16 years old. I told my dad that I was 17, but he knew better and wouldn't sign my papers. My mother wouldn't sign them either. While visiting dad in Toledo one time I met a Coast Guard recruiter. He told me how the Coast Guard cutters escorted Liberty ships across the Atlantic and fought with German submarines. It sounded like there was plenty of action for me. I asked my dad about signing up for the Coast Guard, but I couldn't talk him into it although he did go talk to a recruiter and talked with him about me. The recruiter told him that I would probably be stationed around the Great Lakes during the war. On my next trip to Toledo I asked him again and he decided to sign my papers. I was sworn into the Coast Guard on 7 July 1942.

After boot camp at Port Huron, Michigan, I was sent to Duluth, Minnesota, where the Coast Guard had taken over the Cascade Hotel. The Coast Guard was guarding the waterfront facilities, shipyards and coal docks on the Duluth waterfront.

I didn't really like the idea of guarding an empty warehouse. One day when I returned to the hotel from guard duty one of the men said, "There is an opening on one of the ore boats for an armed guard. Would anyone be interested?" I shouted out, "I would!" So I got my gear together and they took me out to the Cleada Snyder, which was an ore boat that was unloading coal in Superior, Wisconsin. There were two of us Coast Guardsmen on the boat. We checked everyone coming on and off the boat. When the ship was in port we had to stand guard duty. The two of us took turns guarding. We made up our own regulations on that. We would sail to Marquette, Michigan, and load iron ore to take to Cleveland, Ohio. We would unload the ore then sail to Toledo to load coal to take to Hancock, Michigan. We repeated this process until mid-November when they laid the ship up for winter.

That winter I asked for and received a transfer to the USS Amaranth, a Coast Guard Buoy Tender that worked the upper Great Lakes. In the early spring we left port to put navigation aides in Lake Superior, northern Lake Michigan, and northern Lake Huron. When winter came we tied up again in Duluth. About half the crew was transferred back to the Cascade Hotel, so it was back to being a guard on the waterfront for me.

One night when I had guard duty there was a bachelor party for one of the crew. I snuck out the back of the hotel and went to the Green Parrot, a really scurvy bar where the party was. I had a few beers and was about an hour late getting back to the hotel. When I walked in they put me in the brig. I was there about a week, then the shore patrol took me to an office building in uptown Duluth to see the "old man." The captain wanted to know what my problem was. I told him, "Sir, I joined this outfit to fight the Japs and Germans, not to guard some damned empty warehouse!" He told me to calm down and then we talked for a good half-hour. He wanted to know how old I was, about my parents, my school and everything about my life. Then the shore patrol took me back to the brig.

After a couple of days in the brig a shore patrolman came and said, "Get your gear together fella, you're shipping out." They sent me to the Alameda, California, Coast Guard Naval Station. After about a week I was sent to Long Beach, California, to board the USS Bisbee (PF-46), a patrol frigate. I was finally aboard a Coast Guard cutter! The ship was in dry dock on Terminal Island between Long Beach and San Pedro, California. We got it ready for a shakedown cruise and headed for San Diego.

We trained by practicing on our own submarines, as our main objective was hunting submarines. We were to escort Navy ships and fight submarines. We went back to Long Beach where we took on ammunition. Everyone went on a 10-day leave, then we headed for the South Pacific. We stopped in Noumea, then we went to Cairns, Australia, where I met an Aussie soldier who took me up to his home town of Babinda on a little narrow-gage railroad. When we returned to Cairns we went to a bar and ended up in jail. They let us out in the morning; I returned to the ship and nobody knew that I'd spent the night in jail.

From Cairns we went to Port Moresby, then we patrolled the northern coast of New Guinea, bombarding Japanese positions for the 4151 Division. General Doe, commander of the 4151 Division, spent some time aboard our ship. One time after taking a shower, I got into a towel fight with a fella and I began to get the worst of it, so I started to run off and ran right into General Doe. He was coming up from the ship service door with two cartons of Camel cigarettes. When I ran into him I knocked the Camels out of his hands, so I bent down and picked them up. When I handed them to him I noticed that he was a two star general! He said, "Excuse me, son," took his Camels and went on. General Doe was aboard our ship several times.

During the invasion of Biak the Bisbee earned the Commendation Medal for excellence in gunnery. Four days prior to the main invasion of Leyte in the Philippines, we bombarded the area and landed 500 Army Rangers on Homohone Island. That afternoon, about 100 enemy planes flew over but couldn't locate us due to heavy cloud cover.

We were at our battle stations for six days beginning on October 20th. One night the Japanese sent an aircraft out to get us. The plane dropped a torpedo which passed just under us, but the plane almost took off our mast as it flew over.

Our ship was in command of the operation. I never found out exactly what a command-ship did, but every ship that went into Leyte Gulf had to go past us. One day as a cruiser passed, a signalman said, "Lowe, MacArthur is coming in on that cruiser." That was about three days after the invasion of Leyte.

On October 24th the Battle of Suriago Straight began. It was the biggest naval battle in world history. We witnessed the battle from our command-ship location. The Japanese lost several battleships and cruisers.

We ran out of meat when we were there in the Philippines. All we had to eat were beans and bread until a commissary officer went ashore and bummed some Spam and dehydrated eggs and potatoes from the Army. That Spam tasted pretty damn good after eating just beans and bread!

We returned to Pearl Harbor for repairs, then headed for Dutch Harbor, Alaska, and spent the winter on escort duty and search-and-rescue missions in the Bering Sea. In the spring we went to Seattle to prepare the ship for lend-lease to Russia. While we were doing the repair work I was scalded by steam and was sent to the Naval Hospital in Seattle. While I was in the hospital the ship left for Alaska without me. Then the atomic bombs were dropped. The day that the Japanese officially surrendered I was given a 30-day leave and went home. I went back to Seattle and was told I had enough points to get out. I was discharged on 1 October 1945. I was still too young to legally buy beer in a bar. Cliff Lowe and Marion, his wife of 62 years, live in Horton, Michigan. They winter in Florida.

# Murph Jack (Tennison) Marrow, LM 2029

*Age 16— US Army*

I was born in Dallas County, Texas on 8 September 1940. My mother was divorced, and I had two half brothers who remained with my father by name, John Calvin Marrow. At the time of my birth, my mother was living in the home of my biological father, John Murph Tennison, and his wife, Zema. My mother worked in "Dad's" grocery stores (Tennison's Grocery, and Little Butch's Grocery, my nickname). I called him "Dad", his wife "Mom", and my mother "Mama". For the first eight years of my life, I lived in a very comfortable stable environment, and I absolutely adored my "Mom & Dad" Tennison.

Three days after my 8th birthday, my world was rocked to its foundation when "Dad" Tennison passed away. As soon as he passed, in spite of "Mom" Tennison's feeble pleas of protest, my mother and I were expelled from the Tennison homeplace with only a small suitcase, and left walking to find some type of new shelter. From the date of "Dad's" passing, my mother and I moved from pillar to post, staying with any relative who would offer us refuge. Mother was very sickly most of the time, but she managed to work long hours at different places, making long walks to and from each day. We never really had a home to call our own, and I could not understand why I was shunned by my own family members. I did not know that I was the product of a love affair between Dad Tennison and my mother. (As a matter of fact, I later found out that I was the only one in the family who did not know the truth, which I finally found out in my thirties when I was approached in a restaurant by a former family member because I looked so much like my "Dad" Tennison.) A year after "Dad's" death, "Mom" Tennison passed away, and a year later, John Calvin Marrow, my other father, was killed. Losing so many "parents" at such a young age was devastating.

During my brief childhood, I attended school as much as possible, but had to work in the cotton patch picking and pulling cotton for my maternal grandfather, wash dishes in a restaurant while standing on a box, and sold doughnuts to factory workers on a crossing bridge and door to door as a youngster just so we could eat; I worked in a produce department, and walked in the early morning darkness to help a wholesale milkman on his route as a young teen. Then I started hanging out with a friend whose family owned a small restaurant and tourist court; I ran the small grill, cooked, cleaned, ran the register, and washed dishes, and waited on customers. They were also in a branch of the gambling cowboy mafia; I acted as a front man while working in the restaurant, warning the gamblers if the police tried to raid the place. I was recruited by the local police chief to set up the gambling operation. This was a bit too much for me to deal with, so I had to leave town.

During the month of November 1956, I visited the various recruiters in the Dallas Main Recruiting Station. I processed for the Marine Corps, took test exams, etc. But on my way to the latrine, I was hailed by a very large Army recruiting sergeant who asked me what I was doing. When I told him that I was processing for the Corps, he said "Good luck, I don't think you are man enough to qualify for this", and he pointed to an Airborne soldier, parachute deployed. I came back from the latrine, and joined the Army, and remained on active duty for 23 years. I then served in the Texas State Guard for 8 years, and am currently in the US Army Retired Reserve.

During my time on active duty in the Regular Army and in the Texas State Guard, I had the distinct honor and privilege of serving on numerous military installations in CONUS, and tours in Germany, Korea, and Viet Nam. I served in the following military categories and specialties:

Military Police Corps. (Company, Battalion and M P Group), Infantry (foot, armored, light and heavy weapons), Military Advisor (MACV), Drill Sergeant (basic), RO.T.C. instructor and S.A.r. (Junior & Senior Division), Aviation ( crew member and Intelligence), U S Army Recruiting Command (San Antonio RM.S., Dallas R M. S. , Denver R.M.S., Wyoming and Nebraska), Fourth Army Recruiting District (Professional Development), and retired from active duty at Ft. Hood, Texas. I served in the following units:

| | |
|---|---|
| Btry-C, 34th FA, BN | Co. E, 17th INF |
| Btry-C, 13th TRNS BN | HHC 4006.08 USAMTC |
| Det. A, 570th MP Co. | HQ 13th A VN BN |
| Det. B, 570th MP Co. | Co. A 502nd A VN BN |
| 52nd MP Co. | USA INST. SAHS (ROTC) |
| 62D MP Platoon | Peacock Military Academy (ROTC) |
| CO D 2d ARB, 58th INF | ADVTM 68 USMACV |
| CO E 4th BN USAMTC | USARMS San Antonio, TX |
| ADVTM 50 USMACV | USARMS Denver, CO |
| USARMS Dallas, TX | HQ First MP Group, TSG |
| SWRRC San Antonio, TX | HQ 102D MP Group, TSG |
| CO A 102D MP BN, TSG | HQ 104th MP BN, TSG |

Looking back over my thirty one plus years of military service, I must admit that I truly loved and enjoyed every assignment. I gained invaluable knowledge from my experiences. I was recognized with many promotions, honors and decorations. Since my retirement from the U S Army, I have been blessed with business accomplishments far beyond my wildest dreams as a child. The U. S. Army and my faith in God gave me all of the tools necessary to over achieve in life, and for this I will remain eternally grateful. One of my proudest accomplishments in life is being a member of a rare and select group of service men and women who enlisted in the Armed Forces as mere children, toughed it out, and immediately became mature men and women in arms.

Murph Jack (Tennison) Marrow, Msc.D, lives in Lancaster, TX with his wife, Claudia.

# Glenn E. McDuffie, LM2614
# "The Kiss"

*Age 15 - U.S. Navy*
*A World War II icon*

To art connoisseurs "The Kiss" refers to the famous sculpture by the renowned French sculptor Auguste Rodin. However, to most Americans of the World War II era, the term brings to mind the iconic picture of a sailor kissing a nurse in New York City's Times Square. That picture is a poignant reminder of the joy every American felt on V-J Day, August 14, 1945, when word came that the war was over.

Among the crowd celebrating in Times Square that day was Alfred Eisenstaedt, a well-known photographer for Life Magazine who was taking pictures of the celebrants. He snapped a picture of a sailor kissing a nurse, but in the excitement he failed to get their names.

Two years and one month before the picture was taken, a 15-year-old boy was sworn into the Navy. Glenn E. McDuffie, LM2614, was born in a mill house in Kannapolis, North Carolina, owned by the textile giant, Cannon Mills, on 3 August 1927. When school was out in the spring of 1943, most of McDuffie's older friends enlisted in the military. With his friends gone and the knowledge that his brother was a prisoner of war (he was captured by the Japanese in the Philippines), McDuffie was anxious to get into the fray. He did what many young men and women did during the war - he changed his birth year and was sworn into the Navy on 13 July 1943 at the age of 15.

After boot camp, McDuffie was sent to gunnery school. A high point for him during gunnery training occurred when he shot and broke the cable holding a target sleeve that was towed behind an aircraft. Upon completion of gunnery school, he was assigned to the Navy's Armed Guard, the unit that provided gunners aboard merchant ships. McDuffie's first ship was the SS Del Rio. The ship picked up magnesium ore from Cuba and took it to the United States. His second ship was the SS Alexander Lillington. The Lillington delivered 500-poupd bombs, small arms, and mustard gas to the troops in Europe. Each time that the ship would dock in New York, McDuffie would go see his girl friend.

When the war in Europe ended, the need for the delivery of munitions also ended. McDuffie was assigned shore duty and immediately took a 30-day leave to visit his family in North Carolina. At the end of his leave, he took a train back to New York City. He was anxious to visit his girl friend who was planning a birthday dinner for him.

As he left the subway a woman said to him, "Sailor, I'm so happy for you!" He replied, "What for?" She said, "The war is over - you can go home now." The news made McDuffie delirious with happiness - not only for himself, but for his brother who was a prisoner of war. He went into the street yelling and jumping for joy. A nurse stood close by with her back to him. As she turned to him smiling with her arms extended, he ran to her

---

*Among the crowd celebrating in Times Square that day was Alfred Eisenstaedt, a well-known photographer for Life Magazine who was taking pictures of the celebrants.*

---

and kissed her just as Eisenstaedt snapped a picture.

It was 20 years later before McDuffie saw the picture. He remarked, "I was more tickled to see my shipmates Jack Holmes and Bob Little in the background than myself. I knew it was me."

McDuffie went on with his life, worked for the Post Office, played semi-pro baseball, dabbled in race cars, and worked with a florist in Houston.

In 1980, Edith Shain decided that it was time to declare her identity as the woman in the famous photograph. She wrote to Eisenstaedt telling him she was the woman. He photographed Edith at age 62 and the story ran in the August 1980 issue of Life, along with the request for the real sailor to step forward. Eleven men came forward, each with a plausible story, but the identity of the sailor and the nurse remained in dispute. Two women besides Edith Shain claim to be the one in the picture.

It is not difficult to see how each of the sailors strongly believe that he is the one in the picture. On that glorious day in 1945, Times Square was filled with many men and women celebrating the end of a terrible war. Undoubtedly, there were numerous sailors kissing every girl on the street. But which one is the one in Eisenstaedt's picture?

Lois Gibson, a forensic artist with the Houston, Texas, police department used her forensic skills to examine the picture. Gibson has done numerous reconstructions of crime victims' faces using only a skull. The 2005 book of Guinness World Records cited her as helping police identify more criminal suspects than any other person.

Gibson carefully studied the bone structure of all 11 men who claimed to be the sailor. She was able to eliminate all but McDuffie. She claims that the sailor is McDuffie beyond a shadow of a doubt. Although Gibson's evidence is very strong, some internet sites maintain that the identity of the sailor remains a mystery.

*(Editors note: this article was developed from a telephone interview with Glenn McDuffie, and from articles in the Houston Chronicle published on 4 August 2007 and 13 September 2007, as well as several internet sites. [On the internet, search for The Kiss Times Square which leads to several sites that discuss the famous kiss.] Louise and Charles Johnson, M1745, of Spring, Texas, read the articles and met McDuffie. Louise Johnson sent copies of the articles to this editor)*

# O. Glenn McKinney, Jr.

*Age 16 - U.S. Army*

I was born on November 10th 1928 and lived in rural southern Illinois until 1942 when my dad, a rural one room school teacher, moved the family to Chicago Hgts. Ill. Where he had found work. He worked three jobs trying to make up for all of the poverty and unemployment he had suffered during the great depression. I as many others of that era, was a product of depravation and need but we kids were happy and didn't know we were poor. Some of the most pleasant memories of my childhood was making our own toys and games, hoops and old tires, swinging on grapevines, pitching washers and going down to the company store to hear the grown-ups talk about WW1 and the Spanish-American war. I even remember seeing some very old Civil War vets.

On Saturday I'd hitchhike to Vienna, Ill., about 10 miles( it seemed 100 ), to go see Johnny Mack brown, Tex Ritter, Gene Autry, Ken Maynard, Gabby Hayes and the rest of the oater stars, do in all the bad guys. The good guys always won back then. For 25 cents we could take in a matinee, have a coke and bag of popcorn, and for another nickel, buy a bag of Bull Durham with a pack of cigarette papers to roll your own. I picked up the tobacco habit at about the age of 12, I kicked it at 32.

I was adventurous and couldn't keep my mind on school stuff, I dreamed of far away places and yearned to travel. So when we moved up north in 1942 I quit school at 14 in the middle of ninth grade. I got my first job spotting pins in a bowling alley. I was a fireman on the C&EI RR at 15. I got my first chance to travel when at 16 I got a job on the Santa Fe RR as a pole lineman and set out to Texas, New Mexico and Colorado. I worked with a couple of soldiers who were assigned to the crew from the Army, we hit the bars together after work, they thought I was 21 as did everyone else. I guess they secretly induced me to enter the military, what I had tried to do by forging my birth certificate to enlist in the navy in 1944 but was unsuccessful as my forgery was discovered. I quit the railroad and returned to Chicago in the first part of 1945. I drove an asphalt truck until finally I decided for patriotic reasons, I had to get into the service. I picked a Monday June 6th 1945 to report to the draft board, I was 16. I told them I had turned 18 Friday the 3rd but that I couldn't make it in and hoped I wasn't in big trouble. They told me it wasn't a crime and gave me the papers, I filled them out and within a couple weeks received my "greetings letter" and was ordered to Chicago for exam and processing. I passed my physical and was sworn in June 2nd. I was offered a 10 day delay in route but I asked for immediate induction and was sent to Ft. Sheridan, Ill. For processing and left there early one morning on a troop train, secretly, not knowing to where until we arrived at Ft. Bragg, N.C. for 17 weeks of basic training. Since VE day was a couple of months earlier, we were training for the invasion of Japan. It was there on August 6th that I was walking guard duty and heard all of the celebration around the post' later to find out the war was over.

After basic I went to Germany on the USS John Errickson, arriving in Bremerhaven Dec. 31,1945. We traveled all during the night and the next day on a unheated train to a replacement depot in Bavaria. From there I was sent to Berlin on a 40 by 8 train, no seats but at least we had a pot-bellied stove for heat. We traveled mostly through the Russian sector and they had out posts along the route. About every 30 miles or so they'd stop the train and have the RTO corporal open the door and greet us with "welcome comrades". Needless to say it took a long time to get to Berlin and the RTO corporal was cussin those "stupid drunken Russians" every time they stopped us and he had to get out and trudge through snow two feet deep.

I was assigned as a replacement in the 78th Inf. Div. 310th FA battalion. You went home for discharge at that time on points, some of the guys leaving at that time were combat troops and had 20 or 30 points. We new arrivals had 2. After the 78th was disbanded the 310th formed B troop of the 16th constabulary squadron. I spent the remainder of 1946 patrolling the autobahn through the Russian sector from Berlin to Helmstet and other constabulary duties around Berlin. Although we were paid in Invasion Marks (a scrip), cigarettes were the main source for money, they brought the equivalent of $150.00 a carton. For about a year this money could be converted to dollars by buying money orders. Then the army finally got wise and created the

Currency Control Book, after that you could only convert your salary, and not your black market earnings into dollars. Before the institution of the CCB some guys sent enough money home to buy a farm.

I left the completely destroyed and razed city of Berlin in late October of 1946. We left Bremerhaven on the Wilson Victory ship the first part of November and took the northern route to save time, big mistake, it took us 14 days to get to New York. We almost capsized in the stormy seas, we listed 42 degrees once and the capsizing point for that ship was 43 degrees. All of the doors were battened down for 10 of the 14 days, not even the crew could go on deck. Ninety nine percent of the 1400 troops that almost died on that ship were bedridden-seasick' stacked seven high on top of each other, on stretched canvas cots with puke everywhere and no one eating. I guess that's why I don't have a hankering to go on a cruise ship.

I was discharged at Ft. Dix, NJ November 23rd 1946 just 13 days after my 18th birthday. I went on to do another two hitches in the army one from May 1949 to November 1952 during the Korean War when I was stationed in Japan with the Army Security Agency. The other hitch was from . October 1956 to October 1962 where I served in th Panama Canal Zone, Ft. Monmouth, NJ and Ft. Ritchie Md. I was discharged in Oct. 62 and went to work at East Coast Relay Station in Frederick MD as a Dept. of Army civilian and from their I was transferred to the secret bomb shelter located at the Greenbrier Hotel in White Sulphur Springs WV which was a congressional relocation bunker during the Cold War. I retired in December 1985.

I courted and married a beautiful Japanese girl in February 1953, unfortunately she passed away in 1977. She bore me three wonderful children and I now have nine grandchildren, all of which I am very proud.

# Doy Miller

*Age 15 - U.S. National Guard*

In the summer of 1938 on August 7 I was born in Portal, a very small farm town in southeast Georgia. When I was around four, we moved to Savannah, Ga. My mother became the breadwinner, when my father was disabled and could not walk. We lived in a two bedroorn one-bath apartment. We had no car, so trolley car, bike or feet was it.

There were six of us in the apartment, mother and daddy plus me and two brothers and one sister. At a very young age of about six years, I started selling peanuts- boiled and roasted for ten-cent a bag or 3 for a quarter. I did this until I was about 12 - 13 years old. My regular route was through the downtown area during the day, nights at the Grayson Stadium when there was a baseball or football game. When there was a concert or wrestling match on Saturday or Sunday, I would be there selling my peanuts in the Big Park. The biggest time for selling peanuts was at the annual, St. Patrick Day Parade on March 17. We would cook and bag all our peanuts, and then we would drain them over an old clawfoot bathtub on a wooden and screenwire rack that my father built.

By the time I was fourteen, I was skipping school to hang out with an older crowd. So when some of the older guys joined the National Guard in 1954, I joined also at fifteen (18 on my paper work). Soon I was skipping meetings, a sergeant came to my home to tell me I would be turned in to the draft board. I told him don't worry about turning me in, I would just volunteer myself. I talked my parents into this by promising to complete my high school diploma and send home money in then form of an allotment. I was told that that I could put up $60.00 a month to send home and they would add to that. Never happened, the allotment of $60.00 went home but would not approve any thing more. During my entire time in the service, I never drew more than $50.00 a month for myself. I did get my GED diploma during this time.

After volunteering for the draft in less than a week later, I was on a bus to Fort Jackson, Columbia. A few days later I was back on a bus this time for Fort Gordon, Ga. for a eight weeks of basic training. From there I was sent to Fort Chaffee, Ark. for more training. Here I received my orders to go to Korea. But meanwhile I landed in the hospital with pneumonia. By the time I was released my unit had already shipped out to Korea. After a month here, I was sent to the East Coast where I was assigned to a Nike missile unit. Things improved this time, my unit traveled by train to Philadelphia, Pa. I spent about a year in that area then was sent to the Boston, Mass. Nike Unit. I Got out of the army while in Boston around Thanksgiving of 1956 then I returned to my hometown of Savannah, Ga. I went to work for the Sugar Refinery and worked there for five years. In March of 1962 enrolled in John Marshall Law School in Atlanta, Ga. Under the G.I. Rehabilitation Bill.

After finishing school, I worked for several years in the income tax and real estate legal business. Though during that time, I continued buying and selling a variety of merchandise, as I did in my younger years of buying and selling peanuts. I bought discount, closeout and liquidation merchandise from truck loads to factory loads of almost anything imaginable, from food, clothing, furniture, gift items and antiques. I had one of the largest flea market, closeout and antique markets in Atlanta for over forty years until 2007. When my wife of 40 years, Lynn moved to Louisiana to be with her father after her mother died. We sold our home and business in Atlanta and cared for her father after he became ill until he died in June 2008. Since then we bought his home and have remained here until now with our daughter Laura, 31. Our other daughter Melana. 38, son-in-law David and eight year old grandson, Joseph live in Niceville, Fla.

# Charles Mills

*Age 14 - Army National Guard, U.S. Navy*

I was born in a little town in Central PA called Northwood. I was one of six children and we were very poor. We all lived in one room, which was my grandparents' garage. We had no water or electricity and had to use oil lamps and a little Karo heater for heat. When I was 5 y/o, my right femur was broken in three places. It wasn't treated right away since my parents didn't have the money for medical bills. They finally had to take me to the ER and I had emergency surgery done, but by that time, osteomyelitis had set in. I almost lost my leg. I walked on crutches for many years. I had always wanted to serve in the military, but I never thought they would take me. While many kids played cowboys, we all played army. When World War II was going on, we used to have to sit in the dark when the air raid sirens would go off. When the war ended, we all took pans and pounded them with sticks. We were very proud of our military and we let everyone know it.

While in school one day, our eighth grade history teacher told us he was just commissioned as a 2nd Lt. in the Army National Guard. He was very proud. Lester Snyder and I waited until class was over and asked if we could join. The Korean War was going on at this time. I had just turned 14 and Lester was already 14.

My teacher said we looked old enough and he would verify our age. No one checked us and I was in a year and a half. When I was injured in a truck accident at summer camp, they found out my age and discharged me. I was very upset. I then joined the Navy, which was easy since I had prior military service, but was only in for 73 days since my leg had not completely healed. I waited about a year and joined the Army. My leg was still bad and after a year, they gave me a medical discharge. I tried to get back into the military, but they would have nothing to do with me until Viet Nam. At that time, the Navy SeaBees were looking for construction people with construction experience. I enlisted into the Navy SeaBees and after training, I was sent to Chu Lai Viet Nam with MCB #8. I served three tours in Viet Nam. While in Viet N am, I was assigned as Petty Officer in charge of the Reactionary Platoon. Our job was anywhere our inclave needed reinforcement. We would cover that area and we pulled all guard duty for our company area. On New Years Eve, 1966, off the coast, one of my centuries spotted blinking lights about 300 yards at sea. It was nearing the Monsoon season and the South China seas were rough. We could see a ship and it was sending out distress signals. It had gone aground on the rocks. The seas were very rough and high waves were pouncing on the rocks. Being an excellent swimmer, I decided to try to reach the ship. I took the rope and tried to swim out to the ship. The rope got very wet and started to pull me under. I had to drop the rope and go back to shore. I then got a land line and wrapped it around my shoulder and tried again. After about an hour and a half, I finally made it to the ship. It was the LST Manoven County. We established communication to shore. The men were all okay and in good health. They were removed at daybreak. Three of my men and I were honored and the commander stated we would be put in for a Navy Lifesaving award, but we never received it. However, we were still proud to have been there and helped. Two days after the Viet Cong blew up a bridge on Highway 1, outside Anton, we were sent to set up bunkers on both sides of the bridge until it was rebuilt. It got a little wild at times with fire fights at night. We were all very lucky. No one was ever wounded or injured.

From Chu Lai, I was transferred to Naval Support at Da Nang for a year and then back to Chu Lai with MCB #6 where I was wounded and put on the hospital ship, Repose. I was then transferred to the Philadelphia Naval Hospital for treatment. I was there for almost a year. When I was released, I was assigned as a body escort for 2 yrs. out of Dover Delaware. From there I went to Keflick, Iceland for 13 months and then to Panama for 5 yrs.

After my tour in Iceland and Panama, I left the Navy SeaBees, after 14+ years. I then joined the Army National Guard, active duty, as an in-service counselor and recruiter. After 31 yrs total service, I retired with a medical discharge as a Sgt. Major. My awards and decorations were the Bronze Star, 2 Purple Hearts, Army Commendation, Army Meritorious Service, Army Achievement, Combat Action Ribbon, Presidential Unit Citation, Good Conduct Medal, Army Reserve Components Achievement Medal, Defense Service Medal, Viet Nam Service Medal, Republic of Viet Nam Gallantry Cross Unit Citation with Palm, Republic of Viet Nam Campaign Medal, Expert Marksmanship Qualification Badge with Rifle Bar and Army National Guard Recruiter Badge.

Upon retirement, I attended the Pennsylvania Police Academy to be certified as a PA police officer. I worked part time in Bellwood Borough,

Gallitzin and Tyrone Borough before becoming chief of police in Snyder Township and Mont Alto. I worked as a police officer for over 20 yrs. This was one of the most rewarding experiences of my life. It gave me a chance to help many young people get their lives together. If they got into a little trouble, I would give them community service in lieu of arresting them. To this day, many young men and women have come up to me and thanked me. I still work for our youth and I manage the American Legion in Greencastle, P A. At our picnic grounds, we have put up soccer fields, football field and camping for scouting. I guess we never quit serving!

All of my brothers served in the military - Clarence, US Army; Bill, Army Airborne and Ronnie, US Marines. My children that have served - Jerry, US Navy & Army; Andrea, Capt. in US Army and Charley, US Army; as well as my father-in-law, SFC Raymond McConnell who served 20 yrs in the US Army and died in an auto accident while on active duty and my son-in-law, Derek Bowser, currently serving the US Navy.

My wife, Jeanie, and I currently reside in Waynesboro, PA. We have 8 children; 14 grandchildren and 3 great grandchildren.

# Thomas E. Moore, LM2168

*Age 15 - United States Army*

I was born on a farm in Pike County, Georgia, on 18 July 1930. The country was in the Great Depression at that time. People thought things could get no worse, but they did. My dad was what people called a "dirt farmer." Since the dirt that he farmed was so rocky and unyielding, he and his brothers made moonshine whiskey on the side just to keep body and soul together.

When I, the youngest of four sons, was 3 years old, my dad was killed in an accident. My mother, who was 31 years old at that time, moved the family to Griffin, Georgia. My brothers were 5, 7, and 9 years of age. My mother had never been engaged in public work, but she found a job in a cotton mill making the huge salary of eight dollars a week. I remember picking up scrap iron and copper, which I would sell to the junk man who paid me from four to ten cents for a day of scrounging.

With the huge salary my mother made, clothing and feeding her family was more than she could do, but she would not give me up to a wonderful couple who wanted to take me as their own. She vowed that we would stay together somehow. With the help of church groups who saw our need, and an uncle and aunt who sacrificed greatly to help us, we survived.

At age 12 I got into some trouble and had to spend 14 months and 12 days in the boys' training school in Milledgeville, Georgia. There I learned to play a clarinet and how to behave myself.

World War II was going on at this time, and I read everything I could find about the war and saw every war movie I could scrape up enough money to see. I thought the military uniform was the greatest thing in the world. So I began to plan to somehow get into a unifonn as soon as I could. At age 15 I had just about lost hope of getting in the service until I was of age. I wanted the Navy, but their recruiters required proof of age. On 12 March 1946 I was walking down Solomon Street in Griffin, Georgia, and saw a friend of mine whose name was Bobby a 'Kelley. He told me that he was leaving on 27

March and I could do what he had done. He had run his age up and registered for the draft. They had asked for no proof of age. So I figured out what my birth date should be and walked the two blocks to the courthouse. I told the head of the draft board that I was 18 years old that day and would like to enlist. I asked if I could go with my friend. He commended me for not waiting the usual 30 days and signed me up.

On 27 March 1946 we left for Fort McPherson, Georgia, to be inducted. On 28 March I was sworn into the United States Army. How proud I feltuntil I saw my reflection in a mirror. I looked nothing like what I had in mind that I would look like - soldier ... no, yardbird ... yes.

I was sent by troop train (more like a cattle car) to North Camp Polk, Louisiana, (now Fort Polk), for basic training. There I learned close order drill, military rules and regulations, and after scrubbing a day room with a toothbrush (after hours), I learned not to smart off to the drill sergeant.

After basic, I drove a truck on a convoy to Fort Dix, New Jersey. After a short time there I was sent home on leave and then reported to Fort Lawton, Washington. Even though the war was over by this time, we still did not know where we were going. Then they gave us new unifonns, new boots and other necessities, and we found ourselves being loaded on the Army transport ship Marine Devil. Two weeks later, after spending most of the time below deck because of bad weather with a diet of Clark candy bars, we docked in Yokohama, Japan. To this day I cannot eat a Clark bar.

In Japan, I served with the 3186th Signal Service Battalion, attached to the 8th Army. We took care of all the communication services in the TokyoYokohama area. After a little over a year I boarded the Army transport ship General A. W Brewster, headed for San Francisco. For a short period there I worked as a prison guard at the Camp Stoneman stockade. While there I decided to remain in the Army, but I wanted to go home for a while. I joined the reserves so I could hold onto my rank. I was now a T/4, three stripes up and a "T" underneath; some called us "Christmas Tree sergeants."

After returning home and seeing my one and only sweetheart (we had been in love since we were 13), I lost interest in the Army and at the ripe old age of 17, we married. This marriage lasted 54 years and produced one son and three daughters. My wife passed away suddenly on 6 October 2002, ending a beautiful marriage.

When the Korean War broke out, I had forgotten that I was a member of the inactive reserve, but Uncle Sam had a better memory than I. I found myself signing for a registered letter telling me what day I would report for duty, and soon was in Fort Jackson, South Carolina, helping process recruits. After three months there, I was sent to Fort Gordon, Georgia, to help train recruits. Our company needed a baker, so I volunteered to go to the Third Army Food Service School in Fort Benning, Georgia. The Army sent my records to Fort Bragg, North Carolina, by mistake, so I did not get paid for three months.

After finishing baker's school I was sent back to Fort Gordon as the company baker. I enjoyed this job until our temporary mess sergeant was recalled to his company. Since I was the ranking man in the kitchen, I was made mess sergeant. I served in this job until my discharge, after which I returned to my home in Griffin, Georgia, where I have lived since then.

Soon after being released from the Army, I became a minister. I am ordained with the Assemblies of God denomination.

A little over a year after my wife died in 2002, I married a wonderful woman named Gwen. Our families had been friends since 1950. I helped conduct her husband's funeral 13 months before my wife died. We are very happy.

The Army was a profitable experience for me, and I am very proud to have been able to serve my country. Thomas Moore lives in Griffin, Georgia.

# Arthur K (Bucky) O'Connor

*Age 15- South Dakota Army National Guard*

In 1990, I had just turned 50 years old when the call-up came for my unit to deploy to Operation Desert Shield/Storm. I had trained for this moment for most of my adult life. My unit was ready and I was ready. Although nobody knew what was in store for us, we knew that this mission was important to our national security. Patriotism flourished in my unit. When President George H. W. Bush announced that America would lead the coalition to remove the Iraq forces from Kuwait and preclude the invasion of Saudi Arabia, we knew we had an important mission.

I was the Executive Officer for the 109th Engineer Group, a command and control unit which was assigned U.S. 7th Corps. The 109th was the headquarters for over 5,000 engineer soldiers, both active duty and reserve.

We landed in Saudi Arabia on January 15, 1991, when we climbed off the Boeing 747 it was hot, the terrain was flat and we saw nothing but sand. The plane departed and there we were, standing on the tarmac with our gear and all we could see was sand, no buildings, no trees, nothing. About an hour later some buses arrived and took us to Al Kobar Towers. Al Kobar Towers held 38,000 coalition troops and 15,000 refugees from Kuwait. Just two days later we started to receive SCUD missile attacks four to six times a day. We stayed at this location for eight days awaiting the arrival of our equipment. When our equipment arrived we went directly to the 7th Engineer Brigade berm, which was a tent city. After receiving our follow-on orders we moved to a place called "no man's land", which was just on the border of Iraq and Saudi Arabia. We were living within a berm in tents, austere conditions for sure, but there were no SCUD attacks. Our mission was to maintain 500 miles of sand roads from Saudi Arabia into Iraq. We also built an airfield for an unmanned aircraft. Oddly enough, 25 Iraqi soldiers tried to surrender to this aircraft as it flew over them on day in February. On February 24, 1991 the coalition forces crossed into Iraq for the 100 hour ground war. After the ground conflict was over, our mission was to help re-deploy the United States troops. We moved from the desert floor back to Al Kobar Towers and constructed 600 wash points to expedite cleaning of military equipment. Our engineers also constructed twenty-five K-Span buildings, which were 25 meters wide and about 100 meters long. We erected fifteen fast tents, which were approximately the same width and length as the K -Span buildings.

On June 6, 1991 we loaded our equipment and personnel for redeployment to the United States. On June 8th we arrived at Fort Carson, Colorado.

This takes me back to the start of this whole thing. On July 28, 1940 I was born on a kitchen table on a farm in Burwell, Nebraska. I was the fifth of six siblings, two girls and four boys. When I was six years of age my mother and father divorced, and we moved to Sturgis,

South Dakota and lived with my grandmother. I tried to get into the National Guard when I was 14 but they wouldn't accept me because my mother wouldn't sign the papers. I even tried signing for her but the Warrant Officer wouldn't accept the forged signature. When I was 15 I was able to join the National Guard, with the aid of my older brother who convinced my mother to sign the papers.

At age 16, while I was a junior in high school, I was in a typing class taking a time test and I flipped the carriage return, to my surprise it never stopped. It landed about two feet in front of the teacher. I never said a word; I just picked up my books and left the room for the principal's office. Needless to say I wasn't in the typing class anymore. That summer, between my junior and senior years I went to an Army Clerk school in Fort Leonard Wood, MO. That is where I learned how to type.

After graduating from high school, and experiencing the 2nd eight week school in Fort Leonard Wood, I decided I should go to college. I went to the University of South Dakota and majored in football and it was here that I met my future wife. After being selected the most valuable player of our football team I finally decided to get an education. So, I changed schools and graduated from Black Hills State University with a teaching degree. I taught high school for nine years, and was the head varsity wrestling coach. In 1972 I became a full time employee of the National Guard. In 1973 an incident occurred when members of the American Indian Movement attacked law enforcement officers in Custer, SD. Several highway patrol and local police were injured in this incident. In the subsequent months I was assigned to teach riot control procedures to all the units of the South Dakota National Guard, the State Highway patrol, and many local law enforcement agencies.

My wonderful wife, Patti, and I just celebrated our 46th anniversary. We have three children, two boys and one girl, plus six grandchildren. We have lived in Rapid City, at the same address, for more than thirty years. After my return I was assigned as the Mobilization Readiness Officer. Later I was assigned as the Director of Information Management promoted to Colonel. A position I held until my retirement in 1994.

# Charley R. Paradise, M2575

*Age 15 - United States Army*

In 1948, at the age of 15, I went to San Antonio to visit my aunt. Her daughter was visiting at the same time. I told her I would give anything in the world to join the Army. She asked my age and asked if my mother would give her permission. I told her to forget that, she wouldn't. She said, "Then we will have to make you 18 years of age instead of 17." She told me to go with her. We bought some ink eradicator, changed the date of birth on my birth certificate, and then drove to the post office.

The recruiting sergeant looked at me, held my birth certificate up against the light (scared me some) and said, "I'll bet a month's pay that you're not a day over D." I didn't say anything and was given the entrance exam. The sergeant called my name Charlie Paradise and said that I had passed. I also passed the physical. I was given orders and a date to catch a flight to Fort Grd, California. Upon arrival at Fort Grd, I was very surprised to find that Blacks and Whites were segregated.

During basic training we had to qualify with at least a marksman rating with the old Springfield rifle. One morning while in the prone position, the first sergeant was yelling at a tall lanky recruit from Tennessee. He said, "Look here next to you is a small soldier firing his weapon unafraid; you ought to be ashamed of yourselfl Now get down and start firing!" I felt proud - and older too.

After basic training, I was told to go to the 4th Infantry Division Band and audition. I was given an extremely difficult march by John Philip Sousa. I didn't even bother to warm up my clarinet. They didn't need clarinet players that bad anyway, so I decided to reveal my age. I was given a general discharge under honorable conditions.

I returned home and went back to high school.

The band leader was very happy to see me and asked me to help drill the band with marching. I spent my entire days at the band hall practicing the clarinet five to six hours a day. I flunked the 10th grade, but I was bound and determined to get into a military band.

At the age of 17, I dropped out of school and again joined the

Army - this time with my mother's pennission. My dad had recently passed away. I was sent to Fort Riley, Kansas to take basic training all over again. After basic, I was sent to the loth Mountain Division Band to audition. I was sent to a room with a clarinet and was told to wann up. I wanned up for about ten minutes and a sergeant came to the door and told me I didn't need to audition; I was admitted.

I was assigned to the band training unit, a 16-week school. I was given a room to practice in each morning until enough men reported in to fill a class unit. Each morning at 8:00 a.m., I would report to the practice room. Now and then, different men would peek in the window at me. I would ignore them and continue to practice from a music- for-clarinet book.

We finally got enough guys to start a class. I recognized the guys who had been: peeking in the window. They asked me why I was not on detail to clean windows and cut grass. I didn't know what to tell them except, "I dunno." After two weeks of school, the master sergeant in charge called me into his office and said, "Soldier, I'm going to give you a test, and if you pass it, I will give you your diploma early." I said, "Okay."

He took me to a room, gave me a pencil and the test, and wished me luck. I completed the test and waved to him through the door window and he took it to grade it. After a while, he came back with my test and explained all my errors. The one I remember is: name three types of mouthpieces. He said, "You put down clarinet, trumpet, and saxophone; they want any of the following: reed, cup, double reed, and blow hole." He explained all the answers to the questions I had missed. He then gave me a new test and a pencil and said, "I'm going to give you a test and if you pass it I will give you your diploma early."

After the test, I gave it to him to grade. He graded it and came back into the testing room where I was waiting. He began yelling at me, "I haven't seen such a fine clarinet player be so damned dumb! Sign this diploma and report to the lOth Division Band. Incidentally, you made a lower score the second time around."

I was assigned to the lOth Division band. I protested, but it didn't help. The next month I was recommended to attend the Navy School of Music in Washington, D. C. I had never been to Washington. I enjoyed it. Each band in the Air Force, Army, and Navy could send two people to the school. My base chose me and a French hom player.

I returned to Fort Riley and a month later I was shipped to Korea with the 2nd Infantry Division. When I arrived in Japan, I was sent to Tokyo for an audition. While I was wanning up my hom, in walked my old first sergeant. He told the commanding officer that he knew me from Fort Riley, Kansas. Consequently, I was assigned to the General Headquarters Band, located three blocks from the Emperor's palace, surrounded by a big, beautiful park.

In 1953, I was sent to Fort Sam Houston in San Antonio for separation.

I went to high school in San Diego, California, under the GJ Bill and in 1954, I enlisted in the Air Force. I took basic training for a third time and stayed in until I retired in 1971. During my tour, I served in Pleiku, Vietnam, in 1969-1970.

After retiring, I attended the University of Texas Pan American and graduated in 2 Yz years. I taught elementary school for 32 years before retiring. I am proud to be a member of the Veterans of Underage Military Service. Charlie Paradise lives in Edinburg, Texas.

# Willie L. Paradise, M2503

*Age 16 - Texas National Guard*

During the Korean Conflict, I had two brothers serving there. I was too young to join, although one brother went in at 15 years of age. I was 16 when the Korean War was over, but I still wanted to join the military. A couple of friends told me that I would have a better chance of sneaking in if I went through the Texas National Guard.

As soon as I could, I went to a Guard meeting in January 1955. The sergeant asked for my papers and birth certificate. I told him that they were all on his desk. A couple of hours later he came into the classroom and said that he couldn't find my birth certificate. I acted surprised and told him angrily that my mother had warned me about losing my papers. I said that my mother was going to be very mad. It was the only copy of an original. Later that night I was sworn in and fitted with uniforms.

I attended classes with my outfit. The one-weekend-a-month meeting was training in the field war games and firing the Ml. My company, the 1 12th Armored Cavalry, went to Fort Hood, Texas, by train for our 2-week summer field training. I enjoyed the training. On the train on our way back home, I was telling some friends that I sure hated to spend the rest of the summer picking cotton. A sergeant overheard me and said that he was taking names of recruits that wanted to volunteer for active duty. 1 told him that 1 wanted to go. A few days later, five of us from the unit were on our way to Camp Chaffee, Arkansas.

There were 250 of us recruits in Dog Company, 46th Battalion. The company was composed almost equally of regular Army and draftees, with 17 National Guardsmen. The basic training was only 10 or 11 weeks, but seemed like a lifetime. 1 loved it. After we graduated, the sergeant asked me what I had decided to do. 1 looked at him and said, "Sarge, if the Army has more men like you, I don't want any part of it, and by the way, I just celebrated my 17th birthday. You couldn't break me." He was furious and stomped away, mumbling-about not breaking a 16 year-old.

I reported back to my home-duty station and was promoted to private first class (E-3). Then I went to a Navy recruiter and asked a lot of questions. Ten days later, I was on my way to the U.S. Naval Training Station in San Diego, California, as a seaman (E-3). Company 513 was formed on 4 October 1955. Our company chief petty officer looked at me and said, “Must be some mistake; is this you?” He showed me my records and asked if I had just come from Army basic training. I said yes. He took me aside and whispered, “I was going to put you in charge of this company, but these guys will beat you to death. You’re too small and skinny.” He told me that I didn’t have to do all the training with the company since it wasn’t much, but I would have to go swimming and do the knot classes. All I did was go to the post office twice a day to get the mail for the guys and at times go with the chief to the post exchange and drink coffee.

After boot camp, I went to my first of four ships that I would serve on. The USS Ford County (LST -772) was in Bremerton, Washington. I served on her for two years as a laundry man and ship serviceman. We did a 10-month cruise to the Far East. We visited Hawaii, two trips to Korea, Okinawa, Philippines, several ports in Japan, and spent Christmas 1956 in Hong Kong. We practiced war games on Iwo Jima, set foot on Muko Jima, and carried troops all over the Far East.

The LST-772 was decommissioned in December 1957. I was now a ship serviceman 3rd class. I was sent to Kwajalein in the Marshall Islands. In 1958, I was part of “Operation Hard Tack,” testing atomic bombs. My next duty station was the USS Tulare, AKA-112. It was much larger than a LST. It was Big! Although it was an AKA (cargo ship) it did the work of an APA (troopship) by carrying troops. We did a lot of “away all boats” maneuvers.

I reenlisted and reported aboard the USS Gilmore (AS-16) at New London, Connecticut. By then I had ulcers and noticed that I was losing my hearing. The doctor said I should not eat spicy foods, but the ship was serving pizza, lasagna, and spicy spaghetti about every other day. I spoke to my division officer and he said that 85 percent of the crew were Italians and that is what they liked. I had been onboard for a year and was told that I couldn’t transfer! I told the doctor.

One Monday morning the ship was going out to sea for a training cruise. The yeoman on duty came to me and said that I had 30 minutes to get off the ship I was to report to the USS Fulton (AS-II) at Charleston, South Carolina. Whoopee-good southern food! I already knew a couple of guys on the Fulton that had been on the Gilmore. My ulcers got so bad that I spent ten day on baby food at Norfolk, Virginia Naval Hospital. It was there that the doctor noticed that I had a hearing problem. He had me tested. The results showed that I was completely deaf in my right ear. After 5 years and 9 months I accepted the idea that my military days were over.

I was awarded 10% disability for my hearing loss.

I did some work around Texas, then moved to Los Angeles, California where I started working for Greyhound Bus Lines as a terminal employee in 1962. In 1964, I transferred to Portland, Oregon. I spent 25 years with the bus company. When they got rid of the union, the bus company went bad.

My son-in-law was stationed on board the USS Hepburn (FF-1055) and was on his way back to the States from Kuwait when he invited me to join the ship at Pearl Harbor and ride the rest of the way to San Diego. I loved the Tiger Cruise. It brought back many memories. My son-in-law is now a Captain in the Naval Reserve.

After leaving the bus company, I worked for the U.S. Postal Service as a city carrier. After 12 years and my military time, I retired at the age of 62. I graduated with an A.A. degree from Mt. Hood Community College.

I met my wife Phyllis in 1961 while she was in the Air Force. We have 3 daughters. She earned a B.S. from Portland State University. After raising our daughters, she worked for several government agencies. She retired from the U.S. Customs Service as an import specialist Recently, I was appointed Vice-Commander for the State of Oregon, V.U.M.S. group.

Willie and Phyllis Paradise reside in Troutdale, Oregon.

# Donald Pendleton,

*16 Years - Army National Guard*

I joined Utah Army National Guard Battery B 213 Armored Field Artillery 8th day of September 1953. The Unit was returning from being deployed to the Korean Action. Some of the members enlistment was up and they needed replacements.

The 11th and 12th Grade Students at the local High Schools were joining. I was in the Dixie High School 11th grade. Some of my friend ask if I would talk with Captain Wendell Motter, Commander and join the National Guard.

I approached Captain Motter, at his home. I told him I was 16 years old but wanted to join the Guard. He said "Don't tell me how old you are, you look 18 to me." I don't remember having to produce a birth certificate or having my parents sign for me. Between the Utah Army National Guard and US Army Reserve, I have 34 Years.

Basic Training was not required for National Guard Recruits when I enlisted, however my 1st year I went to Artillery Track Maintenance School at Ft. Sill, Oklahoma. The next year I went to Food Service School at Ft. Ord, California, along with annual training at Camp W. G. Williams, Utah.

Then they made Basic Training available, so I went back to Ft. Ord, Calif. Since I had been there a year earlier I knew my way around and some of the instructors at the Food Service School.

They started the Duty Roster at A & Z and worked to the O's & P's. At the end of the 4th week they allowed us to go to town. Paul Milligan, from the California National Guard, my trainee platoon Sergeant and I did a little too much celebrating. As we were driving across post I mentioned that I was not NG but CID doing some investigating of incidents involving trainees in RFA (Reserve Forces Act) Units.

I never said another word as if I had caught myself talking out of school.

I never pulled KP, Guard Duty, didn't qualify the known distance rifle range, the reason being that if it was suspected that those pulling and scoring the target hit was purposely miss scoring the shooter. I never went thru the infiltration course or anything else trainees are required to.

As we were leaving Ft. Ord, Paul ask if I was CID or NG and then related our 4th week conversation. I think that is what brought it back to mind and reinforced it to where I remembered so well.

After Ft. Ord, I went to Ft. Chaffee, Arkansas for Field Wire Communication School.

Later on I went to Ft. McClellan, Alabama to Chemical, Biological, Radiology School before it was changed from CBR to NBC.

Then I changed to Special Forces and went to Ft. Benning, Georgia to Jump School. Since this was a NG Unit and drilled once a week and I was working out of the area I went into the Army Reserve and then back into Utah Army National Guard Battery B 2 Battalion 222 Field Artillery.

Other Forts or Camps I spent time at was; Camp Guernsey, Wyoming, Camp Blanding, Florida, Ft. Lewis, Washington and Dugway Proving Grounds, Utah.

I spent the last years of my enlistment as driver for at least 9 Battery Commanders before retiring in 1987.

I saw no combat or served overseas.

# William Price

*Age 15 Years*

I was fifteen years old when I joined the U. S. Army National Guard of Mississippi, 12 August, 1955. I was assigned as a M-2 tracker operator with A Battery of the 115th AAA at Meridian, Mississippi. Within two years I was promoted to Corporal and assigned in the Arms Room as an armorer. It was there I learned about every weapon from the 45 cal. pistol to the 50 cal. machine gun.

On August 13, 1957 after finishing high school I requested active duty with the U. S. Army and went through basic training at Fort Chaffee, Arkansas. After that eight weeks I was sent to Fort Leonard Wood, Missouri for Combat Engineer training. It was there I requested Airborn school at Fort Benning, Georgia. That was not too bad so I requested Ranger and Pathfinders school. After going through all that I felt I was fifteen feet tall and bullet proof. I learned very soon some of the bar tenders at Phoenix City, Alabama could stack rangers like cord wood to be removed by the local police and escorted back across the river.

The next spring I was sent to Korea and assigned to the 74th Engineer Battalion to watch the K.S.C.s crush rock and stack them into walls. When the 74th was deactivated I was sent to 547th Floating Bridge Company to watch K.S.C.s and KATUSA s maintain the largest floating bridge in the world; across the Hahn River. It was there I was injured in an accident and spent twenty days in the 121 Evac. Hospital at Inchon.

There was a male nurse who was on my ward and I became very interested in his work. I made a decision to change my way of living and become a nurse ... 1 have to say my Dad thought I was nuts.

I was discharged as a Sergeant on 12 August, 1959. and returned to Meridian, Mississippi. Again I joined the Mississippi National Guard to watch helicopters torn apart and put back together.

I needed a job so I went to Anderson's Hospital and ask to become an orderly. There I met one of the nurses who informed me of a school of nursing at Birmingham that took male students. It was here I also met another nurse who became my wife, and blessed me with two fine boys and a beautiful girl.

I had to take a lot of tests and after passing them I was accepted to nursing school at Carraway Methodist Hospital and Birmingham Southern College. Going back to school was a trip, but I finally got the swing of it.

While in Birmingham, I joined the famous "Dixie Division" Alabama National Guard and assigned to the Medical Detachment, even though I was lacking in medical knowlege.

Upon graduating from nursing school I applied for a reserve commission in the Army Nurse Corps and requested active duty. They grabbed me up and sent me to the Officers Basic Course at Fort Sam Houston, Texas. Since I had prior active duty as a NCO, I was made Company Commander of the class. Here I am a Second Lieutenant with all these badges on my uniform. This caused a lot of concern with some of the older officers; so I had to carry my DO 214 with me to ward off all the nosey ones.

I requested to go to the Anesthesiology Course for Army Nurses, and was sent to Madigan General Hospital at Tacoma, Washington. There I was placed in the Operating Room and trained OJT as an Operating Room Nurse.

I came to work one morning was relieved of my duties and sent to Fort Lewis, Washington for some "Top Secret" assignment. I was placed on leave and told to secure my family elsewhere. After returning from leave I was assigned to the 3rd Field Hospital to go where else, Viet Nam.

I was assigned in the operating room. We were not too busy at first but soon we were found out and many times a forty-eight hour day was common place. After about six months we were blessed with personnel from two other hospital and welcomed the relief.

Soon afterwards, some of us with combat wounded experience were reassigned to the 3rd Surgical Hospital. It did not have any female nurses, and soon found out why. We were in tents beside an old French runway in the middle of the jungle. It was not a place I wanted to be; and as faith would have it, I was wounded one evening by someone who did not think we should be there. I was not hurt to bad so I stayed there till I was sent home.

I was assigned at Fort Bragg, North Carolina with Womack Army Hospital and the 82nd Airborn Division. I liked that because I was drawing "Jump Pay". While there I attended the Special Forces Warfare School, and got my Masters Degree from the University of North Carolina. I was volunteered to return to Viet Nam TOY for six month stints a couple times. I also was given the privilege of serving in the Dominican Republic.

I again requested to attend the Anesthesiology Course and this time it was approved. Off to San Antonio at Brooke Army Medical Center for eighteen months. I will have to say I met some of the finest officers and enlisted men while in school. I was not out of Anesthesia School a day when I sent to the Officers Advance Course for another six months in that awful place of beautiful San Antonio, Texas.This was toooo good to be true.

My next stop was Fort Gorden, Georgia assigned to the Anesthesia Department of the U.S. Army Hospital. This was not to last long because the need of anesthetist in Viet Nam was ever reminding. And guess who was asked to go back for another tour? Who else ...

I arrived in Saigon again on 17 October 1970 and was assigned to the 24th Evacuation Hospital at Long Bien. This was a busy place, and I put a sign behind my anesthesia machine "We dose but never close". About four months in the tour I was assigned to the 1st Cav. as a medical officer. It was there I received the Combat Medics Badge and another Purple Heart. While I was recovering in the hospital I was reassigned to the 95th Evacuation Hospital at DeNang. I finished out my tour there. I was presented the Silver Star Medal, Bronze Star Medal and the Army Commendation Medal before leaving.

Returning to the States I was assigned to Walter Reed Hospital. My wife did not want to live in Washington DC with three little kids, so I requested to be relieved from active duty.

I stayed in the Army Reserve and practiced anesthesia, at home in Meridian, Philidelphia and Oxford, Mississippi. While in Oxford I attended Law School at "Ole Miss".

It was here when my life came crashing down, for I lost my wife and daughter in an automobile accident. If that was not enough, alcohol abuse was taking its toll. And as any good alcoholic knows, there was not anything wrong with me.

I stayed in the Army Reserve, moving around the country from one unit to another.

I met a wonderful lady with three small girls in Alexandria, Louisiana. We were married and I have to say she saved my life.

We were living in Phoenix City, Alabama when I was called back to active duty to give anesthesia at Fort Benning during Operation Desert Storm. I was on active duty for eight months this time and returning to the active reserve until I retired in 1994.

I have the pleasure of entering the military as a Private and retiring as a Colonel, obtaining a great education and profession.

# LET'S HELP THEM FIND THE AMERICAN DREAM

Jack D. Harris

Love points the way to treasures untold,
A word fitly spoken is an apple of gold,
Let's give our Vets our deepest esteem,
And help them find the American Dream.

"Dear folks," was the way the letter began
He'd joined the Marines, He wanted to become a man.
For the past several months, no letter had arrived
The parents hadn't known if their boy was dead or alive,
They were so excited to hear from their son
They hoped his hitch was over and his work was done,
With trembling voice, his Mom slowly read each word
She wanted to make sure that each line was clearly heard.

"Mom, Dad, I'm sorry that I haven't taken time to write
But most of the time, we were caught in a fight,
There was crying and dying, O how we feared the bomb
So many of my friends lost their lives in Viet Nam,
"I'm coming home, and I'm bringing a friend
He has no one else, his injuries to tend,
His needs are such that most simply could not meet
Without our help, his heart will no longer beat.

"He saved his men that night when he was on guard
But now, he's almost blind and his face is badly scarred,
He has no left arm and his feet were both blown away
It's only by the grace of God that he lived to see another day,
"I know that what I'm asking you will come as a shock
But my friend can't handle another hard knock,
If you can find it in your hearts to share your love
Surely God in His mercy will bless you from above.

The mother and dad were filled with dismay
They couldn't believe their son was speaking this way,
"Son, they wrote, surely you understand
We're too busy to have a blind cripple on our hand,
"Wish him well and leave him and come on back home
There are many other places where he can roam,
You know we have such great plans for you
Where we want you to go and what we want you to do.

The driver didn't see soon enough to apply his brake
The wheelchair was crushed, the body, no one could wake,
"It seemed that he wanted to die," someone cried
As they turned the lifeless soldier over on his side,
He was missing his left arm and both of his feet
A letter fell from his shirt as they wrapped him in a sheet,
The blood-soaked, tear-stained page read, "Son, surely you understand
We're too busy to have a blind cripple on our hand."

Since we were not there to see what they saw,
the specter of death with its face, wild and raw
If we withhold mercy when it's ours to give,
our days will be bitter as long as we live!
Love points the way to treasures untold,
A word fitly spoken is an apple of gold,
Let's give our Vets our deepest esteem,
and help them find the American Dream.

# Ralph J. "Jake" Raby, LM1156

*Age 15 - United States Navy*

I was born in Gastonia, North Carolina, on 26 November 1929. Life was tough in a mill town. Our family had five children - four boys and one girl. We lived in a small house near the mill where my parents worked. Myrtle School on Myrtles School Road in Gastonia was a few blocks from our home. There I finished the fifth grade before going into the service.

Although my childhood was short, my brothers and sister and I would go fishing, hunting, and worked in the garden during the spring, summer and fall. During the short winter months we would help Mom keep the coal stove fed, go to school, and wait for spring. Mill children did not have a lot of opportunities in life. They were enticed to drop out of school and go to work in the mills. I saw the service as an opportunity to serve my country, see the world, and have a better life.

My father agreed, and in October 1945 my dad and I went uptown and he did a little finagling at the courthouse. He put my name in front of my brother Herman's age on his birth certificate, which made me old enough to join up. My dad told the clerk at the courthouse that they had our birth dates wrong, so she switched them. It was easy to get things done that way in the early 1940s, as all documentation was entered into the record books by hand.

I enlisted in the Navy on 10 October 1945 and was sent to boot camp at Camp Perry, Virginia, just outside of Williamsburg. It was a Seabee base. I trained with the 1903 Springfield bolt action rifle. After boot camp, I was assigned to the USS Kishwaukee (AOG-9), a tanker. We sailed to Japan and on to Korea with a cargo of aviation fuel. We passed through many mine fields on the way to Korea, which made the trip on a tanker full of aviation fuel very treacherous.

A year and a half after I joined the Navy, my brother Herman joined the Navy as well.

Thus, the ruse about my age was discovered. At this point, things became a little complicated. I had run away from home to join the Navy with the help of my dad, and I did not want Mom to know where I had gone. Of course she knew that I was in the Navy, as she was determined to find out where I was stationed. She enlisted the help of the Red Cross to find me.

The Red Cross contacted my commanding officer, the ship's captain. He was due to go home the next day and did not want to get involved in a situation that might delay his discharge. He called me in and showed me the telegram that revealed my true age. He asked me what I wanted him to do about it. I told him I didn't want him to do anything because I wanted to stay in the service. I was having a good time in the military. Why, they served the best breakfast I'd ever eaten. I'd been used to having fatback, gravy, and biscuits in the morning. God, the Navy served bacon, eggs, taters, grits, butter, jelly, coffee, and milk! I sure didn't want to give that up. The commander tore the telegram into shreds and tossed it into the trash can. Then he gave me a direct order to go upstairs and write my mom a letter. I did write the letter and everyone was happy.

I was discharged from the Navy on 31 December 1948. When the Korean War broke out in 1950, I joined the Air Force and served for four years. In Korea, I was attached to the 452nd Light Bomb Wing, and served at the K9 airstrip near Pusan, Korea.

Many of my relatives were in the service during the same time. I found my brother Bert in a Swedish hospital in Pusan in 1951. He was suffering from frostbite on his feet.

So many children in Korea were orphans. Each soldier tried in their own way to help these living casualties of war. I found one such child wandering the streets of Pusan and befriended him. He was 12 years old, and alone in a war-torn country. He could have gone to an orphanage, but even the children in the orphanages were starving. I think they all wanted to be attached to a soldier who would share food with them, so I adopted him, so to speak.

I sent money home to my relatives and asked that they gather clothing and supplies for him and to ship them to Korea. While my relatives in the States were gathering children's clothing and supplies, I ended up in the hospital for a short time. The little boy was sent to an orphanage. When I left the hospital, I picked up the box that my family had sent, but I had a hard time locating the little boy. At last I found the orphanage where he had been sent. I took the box of clothing, supplies, and toys, to him. Without even thinking of himself, he began unpacking the box and sharing with the other children. It was great to see such an unselfish child in the midst of so much despair and need. All of his little friends in the orphanage got a present that day. Little tee-shirts with Roy Rogers on them and many other American characters that perhaps had no significance to them in the way it did to American children. It had a stronger meaning to these orphans - warm, clean clothing sent from someone who cared for their wellbeing. This was Christmas 1951. I wanted to adopt the little boy, but it was not allowed because I was not married at the time.

After I was discharged from the Air Force, I returned to Gastonia and went to work at the Westvaco plant that manufactured cardboard boxes. I had a long 37 year career at Westvaco as a press operator and later in maintenance. I have been a member of the VFW in Gastonia for many years and served as commander of Post 9337 for 15 years. I will not be seeking re-election a post commander, but will retire to the ranks of VFW membership and a member of the "Last Man's" club.

I married soon after returning to the Gastonia and have children of my own. My wife Bobbie died in 1997, and now it is just me, my son Jamie, and stepdaughter Toni. I sometimes wonder what happened to my little orphan with the sharing heart.

Ralph Raby lives in Gastonia, North Carolina.

# Jerry E. Ragan

*Age 15 - U.S. National Guard*

I was born August 26, 1934 in Burlington, Co in a cabin just off the main street. My parents were young and divorced after a younger sister was born in Denver and we starated living with a great aunt in Burlingon. Our mother stayed in Denver and father did too until he entered the Army Air Corps at the start of WW II. I was back and forth between Denver schools and Burlington, with one year in Muskogee OK and another in Del Ray, CA. Then 3 rural schools outside Burlington, then living with family friends, then an aunt and uncle. My sister was there at aunt and uncle's on fathers side. Then mother took us to Denver where I lived in a home for boys, attending Byers Jr. High and my sister with mother in a different Denver area. The boys home moved in to a camp near Bailey, Co for the summer each year. I was there over 2 years when our father returned and took me to his fathers home in Burlington for High School.

I had always been a top student and interested in music, but my freshman year I flunked every thing. I was in band and vocal performanace and stayed in school, but was shy many credits, but kept up for sports. Friends talked a buddy and I into joining the local national guard at 15 years old, Infantry Co I, 157th Reg. Combat Team, the Armory was a focal point in town.

We went to summer camp at camp Carson, Co and the first year joined the regulars in fighting remains of wildfires on and off post. This was in 1950. I was the B.A.R. man, we only had one, and was made the trainer on that weapon in the unit. The next year the 1st Sgt said I would be training soldiers at Carson in summer camp with the regulars on the B.A.R. range and for authority purposes; I was made "acting corporal". During that year I scored highest in the unit with M-l rifle. B.A.R. and other weapons for familiarization, even outscoring the CO members who went to WW II as a unit and stayed in after. As a musician and lead drummer in School I had no problems marching and drilling others and during that years annual I.G. inspection made an impression and was spot-promoted to staff sgt. by a general conducting the inspection. I quit school in early spring, drove dump truck, and after wheat harvest and summer guard camp went traveling on a sales crew. Left that in Iowa, finding a relative of grand dads there, turing 18 and began work in the Maytag factory. I returned to Co during mothers illness and after she passed. After some time, I married and went back to Iowa. I later joined the Iowa N.G. in Newton, Ia and was a prime-mover driver in an artillery unit.

# Kenneth E. Rogers, LM1676

*Age 16 - United States Navy*

I was born in Biloxi, Mississippi, on 8 May 1926, the second son of George and Della Rogers. During the Depression years, growing up on a farm meant working hard to help feed a family of eight. When the United States entered World War II, my first cousin and I made plans to enlist in the Navy. The trouble with this plan was that we were both underage and our fathers' signatures would be required. Both fathers declined to accept our plan.

At a later date my cousin's father did sign and he bid me a sad farewell. I convinced my father that I really wanted to go, so he agreed to sign if the recruiter would also sign a statement that he knew that I was underage. For some unknown reason, the recruiter agreed. I believe he had a quota to fill and the train was ready to pull out.

I was sworn into the Navy on 3 March 1943.

That same day my father drove me to the train station. Before I boarded the train he asked if I had any money. I replied, "No," so he gave me two quarters. It was three years before I saw any of my family again.

After completing boot camp training at San Diego, California, I attended landing-craft school. After that, on my 17th birthday, I went aboard a Dutch troop transport headed for the South Pacific. I was later transferred to the USS *John Penn* (PA26). The *Penn* was sunk at Guadalcanal and I was reported missing. After the Navy found me I was put aboard the USS LST-339. This was my home for more than two years.

My job aboard the 339 was landing-craft coxswain. I took part in 22 landings, such as the Solomon Islands, New Guinea, Philippine Islands, and others. The LST-339 received several awards and commendations such as the Navy Unit Commendation and the Philippine President Liberation Commendation Riggon.

After the war I returned home and was discharged in February 1946.

On 9 March 1946 I married Betty Lou (McAllister) Rogers. Very shortly after that I went back in the Navy and completed 20 years. I transferred to the Fleet Reserves on 10 August 1964 and with Betty Lou and our four sons, we moved to Pensacola, Florida.

I feel very blessed that after four wars, including the cold war, I still have two legs and two arms as well as most other needed body parts. I am humbled when knowing that others were not so lucky.

Ken Rogers lives in Pensacola, Florida.

# John F. Rutledge, LM1422

*Age 15 - Illinois National Guard Age 16 - United States Army*

I was born in Chicago Illinois, on 8 August 1931. My military life began at the age of 14 when I enrolled for high school ROTC during my sophomore year.

The Illinois Reserve Militia was organized to replace the Illinois National Guard (33rd Infantry Division) that had been federalized and shipped to the Pacific theater when World War II broke out. I don't recall if I was asked about my age, but I was 14 when I signed up.

Most members were either World War I veterans or young guys like me. I weighed about 120 pounds, and it wasn't easy training with an old 1917 Enfield bolt-action rifle with an 18-inch bayonet. One weekend we fired on the range at Fort Sheridan, Illinois. I still have my marksmanship badge for firing expert with the Thompson 45-caliber submachine gun.

The National Guard returned in late 1945 and I joined early in 1946 at the age of 15. They were hard up for recruits and needed a certain number to activate each company. I remember being sworn in at least six times in one night. I attended summer camp and range firing at Camp Logan, near Chicago, in 1946.

In September 1947, at the age of 16 years and 1 month, I enlisted in the U. S. Army, using my brother's birth certificate, changing the name. The woman at the reception center looked at it very suspiciously and destroyed it immediately. Several years later while at a service school, I visited the personnel office and told them that they had the wrong date of birth on my military records. I was told to bring in my birth certificate, which I did, and the records were changed. I never heard about this again. However, I did see an army regulation regarding minority enlistments. Basically it said that, if a solder enlisted below legal age and stayed in without a break in service, all of his service counted towards retirement.

I took basic training at Fort Ord, California, from September to December 1947. After basic I was sent to the 5th Infantry Division NCO school at Fort Jackson, South Carolina. Then I served as cadre in basic infantry training units in the 11th Regiment. When the 3rd Infantry Division was reactivated at Fort Benning, Georgia, in January 1949. I was transferred there. Initially I served with division headquarters G-3 (operations) section, then transferred to the 15lh Infantry Regiment.

We participated in the major maneuver "Operation Portex" in the Caribbean, on the island of Vieques in January-February 1950, after taking training at the Navy Amphibious School at Little Creek, Virginia. My unit made an amphibious landing on a blustery morning in rising seas. Many injuries were incurred, mostly legs crushed between landing boats and the ship.

The Korean "police action" broke out in June 1950. As the situation got worse, we awaited orders to the Far East. We were alerted in late July 1950 and the division shipped to Camp Stoneman, California, in early August. We went by troopship to Kokura, Japan, after being held aboard ship at sea for a few days. We learned later that we were a floating reserve for the Inchon invasion, and would have been landed there if it hadn't gone as well as it did.

After landing at Kokura, we went by train to Camp Chickamauga at Bcppu, Japan. We trained constantly, taking 10-mile hikes daily up the mountain and back and had lots of weapons training.

Two months later we loaded aboard ships at Kokura and sailed to Wonsan, North Korea, landing there on 13 November 1950. Our commander was General Douglas MacArthur. My company was quartered in an old schoolhouse.At this time the North Korean Army was pretty well beat, and we all expected to be home by Christmas. However, we were suddenly notified by radio that hordes of Chinese were pouring down from the Manchurian border. Our first order was to fortify our building with sandbags and prepare to fight at that location, as the command did not believe that all units could be evacuated with present transportation resources.

We sandbagged all doors, windows, and the flat roof, and prepared to fight. We increased our ammo supply and set up machine-gun positions on the roof. A short time later we were told to burn all of our petroleum and food supplies and proceed to the port area. After doing this, we were loaded aboard a cargo ship, the USS Iroquois. Some of us were on orders to be transferred to the 7th Infantry Regiment. However, all units were in motion and we couldn't get to our new units.

We expected to be headed to the South Korean port of Pusan;

however, our ship headed north and dropped anchor in the port of Hungnam. Our 7th Regiment was in position on shore, stretched over several miles to hold a corridor open for the 1st Marine Division to move to the port from their positions far north near the Manchurian border at the Chosen Reservoir.

I was assigned as a machine gunner on a Navy LCVP (Landing Craft Vehicle/Personnel). We went back and forth from shore to the various ships in the harbor, transporting Army and Marine personnel, many of whom were wounded. As this operation was terminating, we received a lot of scattered small-arms fire from the shore, and our craft was hit several times.

We eventually pulled out and arrived at the port of Pusan at the southern tip of the peninsula. I did get to my unit in the 7th Infantry, and we went into reserve to regain our strength and prepare to move forward. The regiment had been hit hard. One company had lost all six of its officers and all but 37 men from a normal strength of 200. In the following months I saw action with both the 7th and the 15th Regiments.

I was rotated back to the States in January 1952. Several varied assignments followed. In 1956, I decided to apply for training at the Army Language School at the Presidio of Monterey, California, (later known as the Defense Language Institute). I studied the Arabic language for an entire year, and was then transferred to Army Intelligence and sent to Germany, after six months training at Fort Holabird, Maryland. Never at any time did I use the Arabic language. In Germany I was sent to school for one year at Oberammergau, and studied the German and Czech languages. Upon completion, I was transferred to Berlin, where I worked intelligence-collection assignments. After the Berlin Wall went up, I was transferred to Frankfurt/Main, and continued work in various assignments, and I also served as an interpreter for a polygraph operator.

After completing a five-year tour in Germany in civilian clothing, I was cross-trained at Fort Holabird, Maryland, to perform duties as a counterintelligence agent. Subsequently I spent a year conducting background investigations in the western Virginia area. I spent much time at the Virginia Military Institute and at Washington and Lee University at Lexington, Virginia.

My next assignment was as an intelligence analyst in a psychological warfare battalion at Fort Bragg, North Carolina. Here I was promoted to master sergeant (E-8) and was assigned as battalion operations sergeant.

In July 1965, I received orders to Vietnam, stopping en route for language training at the Defense Language Institute. I arrived in Vietnam in late December 1965 and was assigned as intelligence advisor to the ARVN 9th Infantry Division, as a member of MACV (Military Assistance Command, Vietnam). I accompanied troops on combat operations, mostly by helicopter, but also by land vehicle and boat. I worked with a POW interrogation team, kept data on Viet Cong order of battle, and forwarded captured enemy material for evaluation. An unpleasant additional assigned duty was reporting and photographing atrocities committed by Viet Cong forces.

I was promoted to sergeant major (E-9) in August 1966. I returned to the States in November of that year and was assigned as G-2 (intelligence) sergeant major for the Air Defense Center at Fort Bliss, Texas. I retired there on 1 October 1967 and was immediately accepted into a Federal Civil Service intern program as a quality assurance specialist (ammunition surveillance), which required two full years of training in all phases of munitions, including nuclear weapons. This position required world-wide mobility.

In the years that followed I served two tours in Germany and served as an instructor of tactical nuclear weapons at the U. S. Army Ammunition Center at Savanna, Illinois. Probably the pinnacle of my civilian career was my assignment as chief of special weapons surveillance at the largest nuclear weapons storage facility in the free world.

I retired from the Civil Service in January 1994 in Washington, D.C., and was hired by a contractor at Rock Island Arsenal, Illinois, to plan the destruction of obsolete, surplus and unserviceable munitions for the U. S. Army. I finally went into full retirement in January 2004. My wife Helga and I were married at Oberursel, Germany, on 1 February 1963. We have two adult sons.

My life has been interesting as well as exciting - but it's not over yet/ John and Helga Rutledge live in East Moline, Illinois.

# George F. Salisbury

*Age 16 - United States .... Air Force*

I was born in Mount Harris, Colorado on 23 February 1933, the youngest of seven boys. My Dad was a Coal Miner. At the end of World War Two our family moved to Farmington, New Mexico. In May of 1949 my brother Bob and I walked away from home and went on a trip, just to see things. We viewed a Ringling Brothers Circus and

liked it so much that we traveled with them to Florida. We each took care of two Elephants and slept with them in their train car. Bob wanted to stay with them but in late July I wanted to try another time at home. By August I had already decided I could not stay at home even though I was only 16 years of age.

I was looking for a way out, and spoke to one of the military recruiters. The Air Force Recruiter said that he was behind on his "sign ups" and would really like to sign me on but would have to have proof of my age. Thinking hard about this I had a brain-storm. Because I was liked by a man named Bernie D. who was part owner of the First National Bank. Bernie was aware of my age, but he also knew something about my family, so when I asked him to sign for me he said this: "If the recruiter would let him, he would sign for me." The recruiter thought a Bank Owner was a good enough person to believe and so he signed me up.

I was sent from Farmington to Albuquerque, New Mexico by Bus and was sworn in at that time. From there I was put on a train to San Antonio, Texas for Boot Camp (sic) at San Antonio, Texas, where I spent 13 weeks. Then I was shipped to Fort Knox, Kentucky for School. Things were fine for a while until one evening at the Enlisted Men's Club I was happily drinking, and as a result don't even remember leaving there but became aware of things about a week later in another Enlisted Men's Club. I was not in Fort Knox, but in Fort Worth, Texas.

Shipped back to Fort Knox, it was decided I should be sent to Fort Warren, Wyoming to learn Ordinance. I still drank constantly but did not "lose myself" again. Fort Warren was fine duty for me. One of the things that fascinated me was the old wooden barracks that had held the Cavalry in the old days. One building still had a sign inside on the second floor which read as follows: "It is a court martial offense to shoot buffalo or Indians from the barracks windows."

In 1950 I heard a rumor that a war was starting and my buddies joined me in thinking it was "Just a sack of bull". We were wrong however as in acouple of days we were put on a Train with nothing but military personnel on board, and we traveled from Wyoming to San Francisco, California where we sat all night in a large lot (at the Docks). Then we were loaded aboard a Liberty Ship. (A liberty ship is pretty small). I was told that there were about 2,200 of us onboard. We departed California and it was a miserable time. We passed Hawaii at night but could see a lot of lights. I was told that a couple of men jumped overboard near there. The ship ran out of fresh water and was low on food so they stopped at Guam to take on supplies and we were able to get off the ship for a couple of days. That was really neat, as that ship was so cramped.

We traveled from Guam to Manila in the Philippines where a part of our people were off -loaded. The loaded us on 6X6 Trucks and transported us to Clark Field. While traveling through Manila we observed that it was still a wreck from the devastation of World War Two. There was still rubble among the buildings and the tallest building we saw was a three story one. That seemed so odd, since the Main Street there was either six or eight lanes wide.

Arriving at Clark Field, I was assigned to 24th Ammo Supply Depot. We refurbished a lot of WW 2 bombs, mortar rounds, grenades and other ammunition, which were put in proper working order and then shipped to Korea. Some of the older and outdated ammunition we destroyed.

Clark Field was good duty as I had the opportunity to go out the Back Gate through the Bomb Dump and went to Mount Pinatubo where I got acquainted with some of the Pigmy tribe of Negritos. They were really neat people in my opinion and they hunted with the bow and arrow with which they could shoot birds out of the air. I heard that Negritos were cannibals during WWII. These people were very small, being about 4 feet to 4 1/2 feet tall, but they were very strong. I could barely bend one of their Bows.

One winter I had the privilege of a trip to Baggio City on northern Luzon. Temperatures were in the 70's at Clark,Field, but Baggio City had about 2 1/2 feet of snow. I had a wonderful week there and returned to Clark Field. Still unable to control my drinking (and to this day I do not remember what I did wrong but it must have been bad because I was sent back to the States early) Upon arriving in San Francisco I was discharged on 29 February 1952. I received a General Discharge under Honorable Conditions.

The return trip to The US was quite comfortable and I got aboard a large Three Stacker Ship in Manila and traveled to the Island of Mindanao where a number of Natives were loaded. We went from there to Guam where several natives were offloaded to work as laborers. Then we continued the trip to Hawaii where we picked up some more people for the US and then on to San Francisco. This return journey was such a different trip from riding that little Tub that took us the other direction.

All of my brothers served in the Air Force or Army Air Forces. My brother Bruce is also a member of VUMS and my next older brother Bob spent a tour in the Air Force but was discharged as underage, even though he was 20 years old at that time. (Another story here about "The boy who was old enough"). My older brother Chuck had Mom sign for him when he turned 17, so he was only a bit older than sixteen, but is not considered a VUMS.

Life is working for me now and I am living in Cortez, Colorado where I enjoy my kids and spoil my grandchildren, and work when I feel like doing it. Like so many of the guys who entered the military early, I enjoyed most of it and I would do it again.

# Robert Allen Salisbury

*The Boy Who Was Old Enough*

Robert Allen Salisbury was born on 11 July 1931 in Hayden, Colorado. His father William B. Salisbury was a coal miner and sometimes ranch hand and his mom, Bertha, took care of the kids. According to the family stories I have heard, Bob never had a birth certificate, though all of his siblings did. Bob was born just a year and two months after his brother Bruce, so the two of them became very close over the years. When we look at old family photos we most always see Bob standing near to Bruce.

In late 1945 Bruce walked away from home and joined the Army at age 15. Bob wanted to follow suite but when he tried to join the military they discovered that he had a hernia and this made him ineligible to enlist. Bob went to work for a circus in Florida and paid for his own surgery so that he became acceptable for the military service. He was inducted on 16 June 1950.

In 1951 he was told that he had to show proof of age with a Birth Certificate "or else". The threat, led to this unsophisticated young man being discharged on December 1951. Bob arrived home angry and disappointed that he was out of the service, but was still eligible for the Draft until released from that commitment when he was beyond draft age.

After Bob died in October 1996 I began to search for his records to see what I coµld do to make sure he would have a military gravestone. After a couple of years of adventure in searching for paperwork I received a document in lieu of a discharge which named him as an Undesirable Discharge and at the bottom of this document was typed the odd entry: Date of Birth N/A

Then I knew that the old family story was told, Bob could not provide the document which did not exist, and was separated from the Air Force for being underage. Even though he was 20 years old when he was sent home the man looked very young.

Bob's eldest brother joined the Air Force in late 1945 when his mother signed for him to enlist at age 17.

Brother Bruce enlisted at age 15 and retired at age 35.

# Charles R. Scarborough, LM2644

*Age 15 - United States Army*

I was born in Oklahoma City, Oklahoma, on 12 March 1925. My family moved to the Texas Panhandle town of Borger in 1926. My father worked for a major oil company in the field of geological exploration. Borger was our home until 1938 when my father's work required him to be moved about once a year. That so disrupted my schooling that by the eighth grade I became a dropout.

My two brothers had joined the Army and were stationed together at a camp in San Antonio, Texas. I went to visit them and stayed several weeks until my mother finally signed my enlistment papers on 16 November 1940. I was 15 years old.

While we were on summer maneuvers in Louisiana in 1941, a notice was posted asking for volunteers for foreign service - destination unknown. A buddy and I signed up and we were sent to Fort Hamilton, New York, and departed for Iceland in September 1941 as advance troops to replace the British forces there. We were to construct and operate facilities for troops of the American 5th Infantry Division who were to follow us in June 1942.

Life was hard and lonely. Our only contact with America and the outside world was a short- wave radio we ordered from Sears Roebuck in Chicago. We could listen to the entire world, but our favorite station was, of all places, Berlin, Germany. It was an English-language station that broadcast 24 hours a day and played all the popular big band music.

On 7 December 1941, I was sergeant of the guard and had just posted my last sentry at 1:00 p.m., Iceland time, when a Glen Miller piece was interrupted for a news flash from Berlin. It was a running account of the attack on Pearl Harbor, how the Japanese planes came over Diamond Head and hit Battleship Rowand Ford Island. We knew nothing about Pearl Harbor. That night we received two air raids by German planes from Norway.

In August 1943, part of my company sailed for England. I had spent 23 months in that cold and miserable outpost and was ready for a change. We went to Plymouth, England, where we developed portable laundry units for the coming invasion. Again, a notice was posted asking for volunteers for the airborne service. The 82nd Airborne Division had taken part in the Sicily Campaign in 1943 and had taken a terrible loss of men. Several others and I signed up and sent to the division for medical and other processing. We all passed the exams. I was assigned to the 307th Airborne Engineer Battalion. Fortunately, I had several MOSs numbers that gave me a choice of assignment. I completed 28 jumps and was assigned to S-4, Division Headquarters for the invasion.

On 6 June, our plane, a C-47, was loaded with medical supplies, radios, ammunition and miscellaneous equipment, plus five staff officers, one master sergeant and two technical sergeants. I was one of the technical sergeants. A freak storm disrupted the flight pattern and our plane and others took evasive action. In doing so, our plane's port wing was hit by another plane, flipping our plane over into a dive, spilling cargo and personnel all over the interior of the aircraft. I was sitting next to the door talking to the crew chief with my static line hooked up. The crew chief grabbed me and bodily threw me out of the plane. I vaguely remember landing about nine miles inland from Utah Beach. My back was severely injured during the landing. I was later picked up by men of the 501st Regiment, 101st Airborne Division. They turned me over to the 4th Infantry Division medical unit and I was sent back to England on 10 June 1944. I was finished as an airborne trooper.

At the time of my back injury, the medical services of the Army were severely tested with thousands of soldiers wounded in France. Having no open wound, I was shuffled down the line to the orthopedic people for their diagnosis and treatment.

Their diagnosis was severe bruising in the area of lumbar vertebrae 3, 4, & 5. Lumbar 4 was slightly shifted to the right, but no fractures. I was removed from jump status. After a month of rehab I was ready to get back into the fight; as only a stupid kid would do.

I could have gone home, but chose to remain in Europe. We had heard the rumors of people going home and then being sent to the Pacific because of their combat experience in Europe.

After a month in the hospital I was as good as new. With considerable help from the 82nd Division Headquarters, I secured a transfer to'the 5th Infantry Division, which was in France at the time. I joined the 5th Division on 28 August 1944 at Le Meriot, France, and was assigned to a weapons platoon of Company F, 2nd Battalion, 11th Regiment. I served in that regiment until April 1945.

By the time we entered Frankfort, Germany, I had trouble Walking. My back injury had been aggravated by months of very heavy combat duty and was pinching the nerves to my legs. I was sent to the 108th General Hospital in Paris, France, and arrived there the day President Roosevelt died. The doctors operated on my back, opening the area for the nerves from my spine. My spine had shifted 4/5ths of a centimeter out of alignment in the lumbar region and was pinching the nerves to my legs. They said the surgery was just a temporary fix, that I would need further surgery. How true that was. I have had three major operations on my back and still suffer back pain daily.

I had planned to make the military a career, but my back injury prevented that. After several months in the hospital, I was sent home in September 1945. During my stay there, about three months was spent doing voluntary service for the hospital staff in helping process patients for shipment home and to other hospitals. I had served more than four years overseas and not one day in the United States during the war.

After my discharge I married and we had three sons. I earned a GED and an associate degree from San Jacinto Junior College. I went to work for Phillips Petroleum in April 1946 and worked for them for 39 years. I retired at age 60 in 1985. My wife and children are all deceased. I live in the country in East Texas, fish and play poker with other old codgers at the VFW post. The Army took a kid and made him a man. I have never regretted a day I served in the U. S. Army for my country.

Charles Scarborough lives in Livingston, Texas.

# Raymond L. Secules, M2593

*Age 16, United States Marine Corps*

I celebrated my 16th birthday on 7 November 1941. Exactly one month later, on 7 December 1941, Pearl Harbor was bombed. I was having breakfast, on a farm outside of New Castle, Pennsylvania. During breakfast, we had the radio on and heard the news of the bombing. On the farm, I worked for my room and board and went to school at Shenago Township High School.

The following day I signed up for the U. S. Marines. I was about to leave the recruiting office when an officer asked my age. I told him I was 17. They informed me that I had to have a birth certificate. I had to hitchhike to Williamsport, Pennsylvania, to my parents' house, but they couldn't find my birth certificate. After a couple of months I convinced my dad to swear that I was 17. Instead of going back to New Castle, I went to the local recruiting office and told them I signed up in New Castle. I flunked the physical, but the recruiter sent me to Philadelphia where I passed the physical, but I had to return to Williamsport to have a tooth filled. I was then sworn in and shipped to Parris Island, South Carolina, where I was assigned to Platoon 624.

I remember three incidents from boot camp.

The first was that I said no to the drill instructor (DI) when I should have said, "Yes sir!" The next thing I remember I was picking myself up off the deck. The second thing happened just after entering boot camp. At an early morning inspection, the DI asked me if I had shaved. I said I have never shaved in my life. The next thing I knew I was under the barracks with another boot who had not shaved either. We had to dry shave each other. I guess you know that I shaved every day after that, except in combat. The third incident happened early one morning. We had very little time to shave, shower and get to the toilet before inspection. The guys wouldn't get off the toilet. The toilet seats were over a very long trough that had saltwater flowing out through it. Since the seats were all taken, I rolled up a bunch of toilet paper, lit it, then put it in the trough. As the guys jumped off the seats, the other guys sat down. The DI called me in his office after inspection. I thought he was going to give me hell. He started to do so, but all of a sudden he and the other DIs started laughing. We did not have that problem again.

At the end of boot camp I was hospitalized with malaria. I also had major lung surgery. While I was in the hospital my boot platoon joined the 21st Marine Regiment and went to the South Pacific. After my time in the hospital I was assigned to a casual company where I was assigned to the brig as a prison chaser. During that time I had to escort a general court-martial prisoner to the naval brig at Portsmouth, New Hampshire.

I was assigned to C Company, 1st Battalion, 23rd Marine Regiment at the time they started forming for combat infantry training. Shortly after I joined the 23rd Marines, I was sent to take the exam for private first class (PFC). It took three days to learn all the weapons the infantry used. A few months later, they started giving PFC stripes to the guys just out of boot camp.

After the 4th Marine Division was formed, we shipped out to Camp Pendleton, California, for more training. While in Camp Pendleton, the first platoon was marked as a raider platoon. We trained at night against some raiders recently returned from the South Pacific. When we left the States, we went directly into combat. We were sent to the Marshall Islands where I was a light machine gunner. After the island was supposedly secured, we set up beach defenses. I had just come off duty and was lying on my poncho looking up at the stars when a Japanese soldier with a bayonet loomed over me. As I rolled aside, he got me, stabbed me in my side. However, he didn't see the machine gunner that relieved me. Need I say more?

After the Marshall Islands campaign, the division returned to Maui. Hawaii, for further training and to prepare for the next invasion, which was Saipan in the Mariana Islands. I was wounded twice by shell explosions on Saipan. The second time I was evacuated to a transport ship. I couldn't see worth a damn. I went back to shore with a couple of Marines who had been wounded.

When we reported to an officer I told him that I was a flamethrower operator. With a flamethrower, I had to knock out caves and guns. We went wherever we were needed. We were sort of freelancers. After Saipan, we went to Tinian where I eventually got caught. An officer couldn't understand why it took three men to operate a flamethrower. I told him that if one of the three was half blind you do! He then put me aboard a hospital ship, the USS Samaritan, where I met a nurse, Helen Caldwell, whose parents lived across the street from my aunt and uncle on Arnold Street in Williamsport, Pennsylvania.

Eventually, I ended up in the Philadelphia Naval Hospital. I got reacquainted with my old girlfriend and got married. I received a medical discharge in 1945.

Ray Secules lives in Williamsport, Pennsylvania.

# Robert B. Seymour, M2516

*Age 16 - U. S. Marine Corps Reserve*

When I was just two months into my 16th year of life, my high school buddy, Bruce Brill, challenged me to join the local Marine Reserve unit in San Diego, California, where I was raised. We went to the reserve unit's recruiting office at the Marine Corps Recruit Depot, San Diego, and the recruiter signed us up even though we were both only 16. The recruiter changed the record to indicate our birth year was a year earlier, making us 17 instantly. He told us that he already had a reserve unit member who was 15, so there was no problem getting us in. This was in February 1955, not much after the end of the Korean War, and the recruiter needed bodies to fill out the ranks. Since neither Bruce nor I had jobs, we needed the money that the unit paid for attending meetings. However, I had other reasons for enlisting.

I had always wanted to join the Marines, as they looked so sharp in their uniforms when I saw them in San Diego. I wanted to be part of a well-known military unit like the Marines. I had read so much about them in books about World War II and the Korean War. I was a war gamer from age 11.

Growing up in San Diego, I got to watch the World War II warships come and go in San Diego Harbor. That started my interest in collecting World War II miniatures, and then into World War II war gaming, which I still do today at age 68.

I was in the 10th grade the year that I joined the Marine Reserves. I was also a member of the U. S. Army Junior ROTC at San Diego High. I enjoyed the ROTC and learned a lot about military life, including how to disassemble and reassemble the Ml rifle.

The Marines looked like such a disciplined force, so I was curious about them and how they trained their troops. Joining the Marines, even though only a reserve unit, offered me a chance to be part of them, and to see how they operated. The opportunity came almost a year earlier than I expected.

In the spring of 1955, when I was still only 16, I saw a notice at the reserve unit which indicated that reserve members could apply to go to regular Marine boot camp that summer, alongside regular active-duty Marine recruits. It was only temporary active-duty for a little more than two months, but I needed a summer job, so I signed up to do it. It would mean full-time military pay.

Training in the reserve unit was easy, and I didn't think boot camp would be all that bad. Well, boot camp turned out to be much more than any regular summer job I ever had! I was fortunate in that I already had some military training in the ROTC and had kept myself physically fit, so it wasn't always that bad, but I suffered along with the other recruits. Everything was always in a rush, like a war was on.

I made squad leader early on, as I displayed leadership qualities despite my short height and youth. I was the second shortest recruit in the platoon at 5 feet 6 inches. My general classification test (GCT) score in boot camp was 125. I was asked to try out for officer's training but turned it down because I was underage, even though they didn't officially know that. As I looked so young, I think my drill instructor suspected it.

About the time I joined the reserve unit, I discovered weight training. I built myselfup from 135 pounds to 145 pounds in six months despite living alone and eating only one meal a day. No one in the boot camp platoon could approach the number of pushups I could do. I did 78 at one time. Later I exceeded that number. I had not kept my rifle clean enough and was told to report to the drill instructor's hut for disciplinary action. His punishment for me was pushups, and he made me continue beyond what I thought was my limit; I did 83.

I left boot camp as a private first class and tried to return to high school. I had contracted a respiratory illness while in boot camp that flared up while I was in school, causing me to miss too much. In December 1955, when I was officially 17, I joined the regular Marine Corps for a 3-year hitch.

I didn't have to do boot camp over again, but my GCT scores were lost and I had to take the tests over again. I scored 134 this time. The average GCT scores were 100 for high school graduates, which I wasn't, and 120 for college graduates. My score was often higher than those above me in rank, including some officers, which put pressure on me to perform at those levels, which I rarely could with only a 10th grade education. My high score became a curse at times!

I was sent to an infantry training regiment at Camp Pendleton where I again became a squad leader. I was still one of the shortest

Marines in the company. After completing infantry training I was assigned to do clerical work for the regiment. My entire 3-year hitch was at Camp Pendleton, 40 miles from my home in San Diego.

Although I did clerical work, we continued to get infantry training once a week and had to keep our M 1 rifles ready for use. I had scored well with the Ml while in boot camp and in the infantry training regiment. In one of my annual rifle qualifying-shootings, I shot expert rifleman all five days. I was asked to be in the base rifle team, but I turned it down because I liked my clerical duties. My duties included working at the IBM center with women Marines.

I made corporal at age 18 and completed my hitch at that rank. I was discharged from the Marines in December 1958 and went on to earn my evening high school diploma. I finally graduated from San Diego State College in 1969. My Marine training helped me stick with college and earn my bachelor's degree and helped me to stabilize my life, like my own "Boy's Town," so to speak. I have many fond memories, even of boot camp!

Bob Seymour lives in Santa Rosa, California.

# Alfred H. Spinks

*Age 14 - U. S. Navy*

Alfred H. Spinks was born November 11,1928, in Ft. Mill, South Carolina to Harvey Noa and Lula Watford Spinks. He was born to a mill family. His Father and Mother both worked in the local cotton mill. Alfred had two older sisters and two younger brothers. They all lived together in a mill home near the cotton mill where his parents worked. Al, or Mack as his Father nicked named him, played baseball and midget football in his younger years. He also worked in his uncle's gas station pumping gasoline. There were other part-time jobs in the movie theater as an usher and a ticket taker. He was also proud of being a Boy Scout and making first-class.

Mack, as he was called as a young boy at home in Ft. Mill, got bored at 14 years of age with school. He was afraid the war would be over before he could sign up to join the Navy, therefore he forged his Mother's name and changed his birthday on the enlistment papers. His Father wasn't very happy when he learned that his son, Mack had joined the Navy; but he didn't interfere with his son's false statement. He thought that if he wanted to be an adult, then he should find out what it was like to be an adult. So, off to the Navy training he went at the age of 14 years on a special day in September 1943. In that special September 1943 thru November 1943 Mack was assigned to a Naval Training (Boot Camp) in Bainbridge, Maryland. After Boot Camp he was transferred to the Naval Station in Norfolk, Virginia where he was Seaman 2nd class. He was assigned to the pre-commissioning training of the USS U.M. Moore (DE 442) as a gunners mate (striker). In 1944 he boarded the USS U.M. Moore, a new ship that was commissioned in Brooklyn. He went on a shakedown cruise to Bermuda, a Naval (Fleet) Training Center. Returning to New York for replenishment, the crew encountered a German submarine in the Atlantic area. They had a 36-hour running battle until relieved by a Hunter-Killer Group comprised of a carrier and destroyer escorts which later captured the submarine. When this encounter was finished they continued to New York for further replenishment. At this time they sailed on again to Norfolk and escorted a tanker to Aruba. Their orders took them through the Panama

Canal and they continued to a rendezvous with Battleship USS Colorado proceeding further to Pearl Harbor. At this time they joined up with the Hunter-Killer Group USS Corregidor, an escort carrier (CVE) and four DE. Al, as he is now being called was injured on the ship, USS U.M. Moore and needed to be transferred to the USS Corregidor for medical attention. This was done via a mail sack on a high line connecting the two ships. At one point the line got very low and his body hit the high seas before he was safely on board the USS Corregidor. He stayed on the carrier for one week until he returned to Pearl Harbor and his home ship, USS U.M. Moore (DE 442). It was then that they sailed to join the 7th Fleet for the Invasion of the Philippines. During the Invasion in the Lingayan Gulf in the Philippines several Japanese aircraft were shot down. One of our aircraft carriers was hit by Kamikaze and burned out completely.

After the crew picked up all of the survivors out of the water and off of the ship, they were ordered to sink the ship. By this time it was well after dark, and being accompanied by an additional destroyer we sank the ship using gunfire and torpedoes. They then proceeded on with the Task Force. The Task Force consisted of aircraft carriers, cruisers, battleships and destroyers. At the completion of the Invasion the beaches were secured and the Task Force started to head back to safe anchorage. But, another Japanese submarine attacked the Task Force. The crew, at this time, had contact with the submarine by tracking their torpedoes and engine noise with their sonar. After a long twelve-hour battle with that submarine the crew was able to destroy it. The Japanese submarine blew up by a secondary explosion, but the force nearly blew the USS U.M. Moore out of the water. This was due to the location of the USS U.M. Moore being directly above the Japanese submarine at the time of the secondary explosion. This was a definite kill of the big mother Japanese submarine. They happily survived that confrontation and proceeded to rejoin the Task Force Group.

It was time now to sail onto the big island of "Mog Mog" that was east of the Philippines. After the replenishment and joining the Task Force the Moore became part of the team for the Invasion of Iwo Jima. They patrolled around the island for shore bombardment and antisubmarine. After the Iwo Jima Invasion the Task Force returned to Mog Mog for further replenishment and regrouping with the Task Force to sail on to Okinawa for another invasion while screening of carriers and picket duty. When Okinawa was secured they proceeded to join with the 3rd Fleet replenishment Task Force for raids on Japan's east coast.

When the war was over in August 1945 and the cease fire was

announced, they USS U.M. Moore continued on with the Task Force of cruisers and destroyers to Sagamai Won. They stayed until September 2,1945,when the surrender was signed. Later the Task Force proceeded to Tokyo Bay for a much needed "Liberty and Recreation". After a relaxing liberty, the USS U.M. Moore went to the northern island of Japan and accepted surrender of the northern Japanese Fleet.

Later the USS U.M. Moore left Tokyo Bay en route to Manilla to form another Task Force comprised of amphibious landing ships with occupation forces for Japan. En route to Tokyo Bay they destroyed numerous floating mines using gunfire to blow them up. The USS U.M. Moore hit one of the mines, but fortunately it didn't detonate. When it was a safe distance from the ship it was detonated.

Arriving in Tokyo Bay, the USS U.M. Moore was released and ordered to return to San Pedro, California via Pearl Harbor to decommission Al's home ship, USS U.M. Moore (DE 442).

Al was discharged in April 1946 at the age of seventeen. He had earned the Awards of the Asiatic-Pacific Combat Area with five battle stars, the Philippine Liberation Award with two battle stars, and the Presidential Unit Citation.

# James E. Taylor

*Age 16 - U.S. Army*

On 21 January 1947 I enlisted in the US Army at Sherman TX 2 months and 2 days prior to my 16th birthday. I was sent to Fort lewis Washington for basic training. My first assignment was with G Troop 5th Cav Regt, 1st Cav Div at Camp Magill Japan. There I served in a machine gun squad working my way up to squad leader and buck sergeant (acting jack). In 1949 I transferred as cadre to the 24th Inf Div and was assigned to Fox Company, 19th Inf Regt at Camp Chickamaugua Beppu Japan as section leader in the Weapons Platoon 60mm mortar section. I remained in this job training the mortar squads that I would go into combat with in Korea early July 1950. On 16 July 1950 I was directing fire for my section from a position on the forward slope of a hill south of the Kum River, Enemy fire cut the wire used to communicate with my section so I went over the hill to my section position and learned we still had some ammo left and was expecting to withdraw soon. I climbed back to the crest of the hill and directed fire on the enemy by hand signals until our ammo was expended. I must have attracted some enemy attention giving hand signals because when I started to crawl around to go down the hill I was hit in the knee and left leg by enemy fire. It must have been a dumb-dumb because it entered my knee and exited my left calf taking my combat boot with it. One of my mortar gunners William Davis NC carried me down the hill to the section position where a medic gave me morphine but didn't have time to dress the wound. Our Company had issued the order to withdraw so I was assisted across the rice paddies by section personnel. Once we reached the main road I was instructed to remain in a ditch with other wounded. I wanted to stay with my section but knew I would be a burden so I waited with other wounded until we were loaded on a truck to be transported back toward Taejon. We learned we were surrounded when the convoy ran into a roadblock at the first bridge and it became necessary for able troops to proceed around the road block and slip thru enemy lines. We the wounded were carried on stretchers to the top of a hill above and beside the convoy and left there with Chaplain Herman Felhoelter and Medical Officer Capt linton Buttrey. On the morning of 17 July 1950 we could see a North Korean patrol approaching our hill from the direction of the Kum River. Chaplain Felhoelter made the rounds to each of the wounded and urged all that could walk to head toward Taejon. Capt Buttrey left and at the urging of Chaplain Felhoelter took the only weapon among us (.38 cal pistol) with him I was still groggy from a night of too much morphine and my left leg was stiff. Chaplain Felhoelter ask if I could walk or crawl and I told him I could crawl if nothing else so he told me to get away as far as I could because he wasn't sure what was about to happen. I removed my right combat boot and started crawling toward Taejon. I was still in earshot when I heard yelling and gunfire. later I learned that the chaplain and all remaining wounded were killed by the North Korean patrol. I credit Chaplain Felhoelter with saving my life. After a mile or so of crawling and hobbling I was limbered up enough where I could walk but my left leg and foot was paralyzed so I wore the skin and flesh off my left toes from dragging it along. I don't know how far I walked but I encountered a jeep and driver sitting at a blown bridge outside Taejon. The driver took me to a waiting ambulance that took me to an aid station in Taejon.

I was sent by train loaded with stretchers to a hospital in Pusan then sent by a English hospital ship to Osaka General Hospital in Japan. From there I was air evacuated to the USA landing in San Antonio TX on the day my enlistment was up. We were greeted by a colonel who welcomed us to Brooks Army Hospital and added the "Truman Year" to our enlistments.

I was released from the hospital in late 1950 and assigned duties including Platoon Sgt, Field First Sgt and First Sgt in C Co 701st Armd Inf Bn 1st Armd Div. Fort Hood, TX until I was discharged in Aug 1951. reenlisted and was sent to Big Delta Alaska as Provost Sgt, then MP Operations NCO Camp Leroy Johnson New Orleans lA, then Stuttgart Sub Area MP Operations NCO Robinson Barracks Stuttgart Germany, then 502 MP Operations NCO Fort Hood, TX, Then MP Operations NCO & Calibration NCO 65th Ord Co Korea, then Nuclear Weapons Electronics School Fort Monmouth NJ, then Nuclear Weapons Electronics Schools Sandia Base NM, Then Nuclear Weapons NCO Fort Hood, TX then Nuclear Weapons Operations NCO AWSCOM Pirmasens Germany, then Nuclear Weapons Operations NCO Fort Hood TX. I retired from the US Army 1 Jun 1968 at Fort Hood TX. Betty my wife of 57 years and I reside in Killeen TX.

# Theodore N. Taylor, LM1526

*Age 14 - United States Navy*

It was the fall of 1942. I was a sophomore at East Liverpool, Ohio, high school, and I was bored to death. The class I detested most was English II. The only good thing about the class was that I sat by a large window which gave me a great view of the Ohio River. I could watch the tow boats with their tow of barges loaded with coal or iron ore work their way up or down the river. Once every couple of weeks, an LST built in Pittsburgh, Pennsylvania, would sail down the river.

I envied the men who manned these ships and wished that there was some way I could get into the Navy, but you had to be 17 years old and have your parents' consent. I was 13 and knew my mother would never sign for me to enter the Navy. I had no interest in nouns, verbs, or adverbs, but the English class gave me time to think. I had learned that if you could prove you were 18 you could get into the Merchant Marine. I thought maybe I could alter my birth certificate.

I bought a bottle of ink eradicator, secretly got my birth certificate from where it was kept, and went to work. My birth year was 1928, but it needed to be 1924. Talk about a flat failure - not only did the eradicator remove the year of my birth, it also made an huge hole in my birth certificate! It was time to come up with plan two.

In the spring of 1943, a friend of mine registered for the draft. He showed me his draft card and explained that all you do is go to the local draft board, give them your birth date, and they would register you. For once, English class was a blessing. I spent an entire period memorizing a new birth date that would make me 18. After school, I walked uptown to the city hall, climbed the stairs to the second floor, and very timidly went into the draft-board office. The secretary hardly looked up as she asked me what I needed. She registered me when I gave her my new birth date. It was that easy - no questions, no birth certificate, no parental consent - that was that; I was registered.

When my questionnaire came, it created a stir between my mother and me. The upshot was she went to an uncle for advice. He told her to let me go and that two weeks in boot camp would be all I could take. I was drafted on 6 September 1943.

The draft I was in included about sixty men from our county. After our physicals were complete, we were all standing naked in a large hall. I was between two Charles Atlas-type men. What a contrast! I stood nearly 6-feet tall, but weighed only 118 pounds. As we awaited our assignments, a Marine sergeant in full dress blues, medals, stripes, and all that, strode into the room. He loudly announced that the Marines needed one more man. He also said that there was only one more man that qualified for the Marines, and that man was named Taylor. He marched in front of the line and stopped at the man to my right. "Are you Taylor?" he asked. The man said, "No." He then went to the man on my left and asked, "Are you Taylor?" The man said, "No." The sergeant looked at me in amazement. His chin dropped and he said, "You don't want to be a Marine do you?" I squeaked out a, "No," and he marched off I think we were both happy because I was going into the Navy.

We reported to the Great Lakes Naval Training Center on 13 September 1943. Boot camp went smoothly for me. When I wrote my first letter home I informed my mother that I loved it and I would probably stay for 20 years. So much for the two-week cruise my uncle had predicted.

After boot camp, I was assigned to signal school at the University of Chicago. The instructor who greeted us started by saying that those who made it out of signal school with top grades would have their choice of duty, but those who flunked or barely made it would go to the amphibious forces. Then he quoted some figures about the number of casualties the amphibs were incurring. I worked hard, and for the first time I wished I had paid more attention in English class. Spelling was a big problem, and we had to spell correctly. Somehow I got through with a respectable grade-point average. When interviewed as to my choice of duty, I asked for fleet duty. So much for promises - I was sent to Little Creek, Virginia, tor amphibious training.

I was assigned to a newly formed crew and we commissioned the LSM-147 in June 1944.

We were assigned as a training ship for a time, then we were sent to the Pacific. When we left the dock at Oakland, a pilot was assigned to our ship. He was to anchor us for the night before we were to leave for Hawaii. We had just passed under the Golden Gate Bridge when the pilot ordered our stern anchor to be dropped. Our captain complained that we were not in an anchor zone, but the pilot insisted. So, the hook was dropped and the pilot left the ship. The next morning we prepared to get underway. Orders were given to hoist the anchor. The stern of the ship began to go down, down, down. 'When the anchor appeared, attached to it "was a long, heavy, lead-coated cable. We had snagged the main telephone line between Oakland and San Francisco! We had another ship close-in and attach a line to the cable. We then dropped the hook while the line held the cable up and we were able to move away. I wonder what the people who were using the phones heard?

We hauled various and sundry items, tanks, trucks, etc., to various islands, including Tulagi in the Solomon Islands. We then went to the Russell Islands where the Marines had a large gasoline storage area. We were loaded with about 1,000 barrels of gasoline. While we were there we received our mail. The captain received orders for me to be returned to the states as soon as possible - my true age had been discovered. When we got to Eniwetolk in the Marshall Islands I was put ashore to await transportation back to the States. As I stood and watched my ship sail. I waved and tears ran down my cheeks. This was the first time since my father had died in the summer of 1942 that I cried.

On the way back to the States on an APA, the news came that the war was over. We docked in San Francisco sometime around the end of August. 1 was sent to Treasure Island "here I was assigned to an underage section. There were perhaps a dozen or so men in the unit. I soon learned some facts: one could receive a bad conduct discharge because we were guilty of perjury, having lied about our ages when we entered the service; we had our cards stamped "minor" in very large letters: we were to receive no mail; and we were not to be paid. I had not been paid for several months. My bankroll consisted of about a dollar in change.

I spent a lot of time on guard duty while at Treasure Island, and a lot of time in interviews with officers. Yes, I was worried about the type of discharge I would be given. I had some time to think. I decided I would go back to school and use the G1 Bill to the fullest. It had not escaped me that officers "were educated" most had at least some college.

1 was sent to Sampson Naval Training Base in New York and "was granted an honorable discharge on 22 September 1945. I was not only given my discharge, but I was also given an escort to the train by a young Marine who gave the conductor the ticket to take me home. You know, that made me feel like s___.

My 17th birthday came on November 1945. Several days later, I received by mail my back pay and my mustering-out pay. My Naval career was ended and by that time I was back in high school.

I went on to college and became a teacher — what else!

Ted Taylor lives in Bradenton, Florida

# William A. Tomolonius, M2540

*Age 16 - United States Navy*

My father was in the U. S, Navy as a boatswain's mate during the 1920s. One of the ships he was on was the USS Arizona. He left the service just before 1 was born on the 29th /30th June 1927, at the stroke of midnight, in South Boston, Massachusetts,

During my early youth 1 would look at the pulp magazines my dad would read, such as Wings at ten cents a copy, and of course Our Navy. From that experience I wanted to fly and be in the Navy. Depression days were hard on us.

On Pearl Harbor Day, I asked my dad about the Japanese, as all the people were saying we would beat them in six months. His answer, since he had observed them in the Orient, was, "They paint their ships black; it will take a little longer." He reenlisted again and served on the LST-312 in support of the assault landings at Oran, North Africa, Gela, Sicily, and a real hard one, Salerno, Italy.

As the war dragged on I could not wait to be 17, so I altered my birth certificate and was sworn into the Navy on 22 February 1944 at 16 years of age. I shipped out to Sampson Training Station, New York, for boot training. Before I completed my training, I was posted to be assigned to the destroyer fleet. I had requested naval aviation and went to check on why I was assigned ship duty. The counselor informed me I had a nose problem (deviated nasal septum), and it prevented me from an air assignment. At that I cried, "What could I do?" I cried! The good man said, "Stop," and said he would send me to the hospital to fix my nose problem, which they did!

I returned to boot training in a company that was made up mostly of boys from Brooklyn, New York. They took good care of me and I learned some new words and phrases. I was assigned to the Naval Air Technical Center, Memphis, Tennessee, and graduated as an aviation ordnanceman third class. From there it was to aerial gunnery school at Yellow River, Florida, then to operational training at Banana River, Florida, in Martin "PBM" seaplanes. I was the air bomber and had responsibilities for five gunners and I was the "buoy catcher" that had to "make them" in order to anchor the aircraft at sea. I was just past the age of 16.

After operational training, our plane crew met in Alameda, California, and flew to Oahu, Hawaii, in just 17 hours. From there we flew across the

Pacific to Leyte in the Philippines and then to Patrol Bombing Squadron 20 (VPB-20), at Tawi Tawi, Southern Philippines. There we conducted air/sea rescue patrols and convoy support for the Australian landings at Balikpan and Brunie Bay in Borneo, which I believe were the last amphibious landings in World War II. Incidentally, both were oil-producing installations.

The war ended as we were operating out of Lingayen Gulf, working with submarines supporting the Air Force bombings of Formosa. I was on a flight crew that was transported to Jensen, Korea, by the seaplane tender USS Currituck. We were hit by a typhoon for three days and nights, with wave heights reaching 40 feet. Back to flying throughout China was welcomed after that.

I left the squadron and was transported to Hawaii on the cruiser USS Miami. From there, on a troop carrier USS Dark (APA-159), to San Francisco. I knew I was home in the USA when I saw the girls' faces meeting us at the landing dock. They looked like painted Indians with their cosmetic coloring and red lipstick. We were used to seeing Asian women without makeup.

After a train ride across the country (five days each way), I was assigned to ordnance crew aboard the aircraft carrier USS Hancock. We cruised to Guam and back to Pearl Harbor. Because of my aircrew status, I was transferred to Utility Squadron One (VU-1) at Ford Island. Our duty included towing targets for ships entering Pearl Harbor. I was the tow-target operator and had the pleasure of being the first to release a target sleeve next to a destroyer for their retrieval, in order for them to see if they had hit it. I always thought this proved the feasibility of destroyers to pick up the Mercury space capsules when they returned to earth.

One day at the field, I was assigned to be duty driver. That's when I learned to drive a vehicle. As I was maturing and thinking as I drove the officers and their wives that the only difference between us was a piece of paper. So, I started to study. I left the Navy to go to college, but I had to get my high school diploma first. Through the GI Bill, I graduated as an engineer and started work designing and testing coal pulverizers for public and private power-generating stations.

Subsequently, I worked for General Electric and Pratt & Whitney Aircraft, designing and testing jet engines and components. I ended up my career at General Dynamics' Electric Boat Division in the capacity of initiating programs for the maintenance, alteration, and modernization of the Navy's submarine fleet. The last class of ships I worked on before I retired was the Trident Class Fleet Ballistic Missile ships. And strangely enough, I had under my supervision a retired captain, commanders and chiefs to assist me. Having that piece of paper was important after all! I fulfilled my dream of flight when I acquired my aircraft pilot's certificate.

Throughout my life in uniform or civilian clothes, I have supported aviation and the Navy. During that time I married an angel, 58 years ago! We have 3 sons and 5 grandchildren. I have been lucky enough to live long enough to enjoy them. However, I still cry - I cry during parades as the flag goes by, and when I think of those young people who never returned from their patrols, I pray that God makes it up to them for their loss.

William Tomolonius lives in Newington, Connecticut.

# General John W. "Jack" Vessey, Jr., LM2584

Age 16- U.S. Army

General John W. "Jack" Vessey, Jr., USA(ret.) recently became a member of the Veterans of Underage Military Service, the second member of the VUMS who achieved the four-star rank. General Vessey, LM2584, began his military career as a 16-year-old private, rose through the ranks and retired after 46 years of service, the last four years as Chairman of the Joint Chiefs of Staff.

The VUMS other four-star member was the late Jeremy Michael Boorda, LCM400, who enlisted in the Navy at age 16 and rose from an apprentice seaman to the rank of admiral. He was the Chief of Naval Operations at the time of his death.

Jack Vessey grew up in the small town of Lakeville, Minnesota, where he was born on 29 June 1922. His dad, a World War I veteran, worked for the Minneapolis Northfield and Southern Railroad. The Great Depression was gripping the nation, causing his family to move to Minneapolis when Jack was in the ninth grade.

While still in high school, he enlisted in the Minnesota National Guard by using a little creative math when he put down his age. This was in May1939. It was pretty clear that Europe was getting ready to blow up, but that is not why he joined the Guard. It was an opportunity for a little adventure, to do something different. Also, they promised him that he would get to ride a motorcycle if he joined up. By the summer of 1940 it was clear that Minnesota's 34th "Red Bull" Division was going to be called up, but the call didn't come until February 1941. The troops knew long in advance that it was going to happen.

Jack had been working as a package wrapper at the Sears and Roebuck company on Lake Street in Minneapolis, making about 25 cents and hour, or some astronomical wage like that. He didn't think there was any use in starting college, knowing that his outfit would be leaving soon.

The day his unit was activated they marched from the Minneapolis

Armory to the Milwaukee Depot. It was 20 degrees below zero. It was a lot warmer at their destination - Camp Claiborne, Louisiana. The 34th Division was not anywhere near wartime strength, so draftees were coming in to fill the ranks. There was no basic training at that time. All training was done inside the unit, so the draftees went directly to the divisions.

Jack was promoted to corporal and changed his military spccialty from a surveyor to a radio operator, mostly because there was more money in it. At a training inspection, the sergeant in charge of the radio unit failed the test, so Jack was asked if he wanted to be in charge of the unit. When he told them that he really didn't know that much about radios, he was handed a Field Manual 24-5 and was told that they were going to give him an exam on Monday morning. They said, "If you pass, you'll be a sergeant. If you fail, you'll be a private." Jack spent all weekend with that manual. On Monday, he managed to pass.

He was in the Headquarters unit of the 59th Field Artillery Brigade. On 7 December 1941, the troops were listening to the news about Pearl Harbor on the radio in the day room. The news came as a shock to everyone. Jack had splurged and gotten a subscription to Time magazine and tried to keep up on world events. If you followed the news, you knew something was coming. It was a shock to him that they bombed Pearl Harbor, but it wasn't a shock that the nation was at war.

On 8 December the 34th Division was sent to the Gulf Coast near New Orleans to guard the beaches. Knowing that an assignment overseas was imminent, the division sent half its men home on leave, and when they came back the other half were sent.

The 34th Division went overseas in early 1942 on the British ship Aquitania as part of a huge convoy. The Germans were sinking Allied shipping in great numbers at that point. The troops watched ships in the convoy disappear. The 34th was the first American division brought over to Europe, and was sent to Northern Ireland. In September 1942, Jack was promoted to first sergeant of his company. At the same time, the Army decided that all first sergeants would be master sergeants, so he got a double jump in pay and rank on the same day. At age 20, he may have been one of the youngest first sergeants in the Army.

As a first sergeant he was tough, but first sergeants had to be tough. The company was loaded with guys from Minnesota, Wisconsin, North Dakota and South Dakota. It was a good outfit. The men were hardworking and they were smart.

The 34th Division took part in the American invasion of North Africa, landing near Oran, moving by truck across Algeria into Tunisia. They arrived in February 1943, just before Rommel's big attack. The division was assigned to the French 19th Corps. The troops were fed into the battle as they arrived, and it was a big mess.

The division did not fare well in the early encounters with the battle-hardened Germans. German tanks were running all over the place, plus the Germans had a lot of air support in those days. The Americans were in all kinds of trouble. It was a lesson for all. The division was well trained, but they were not trained for the war they came to fight. It was a lot harder than the leaders had envisioned. They were miserably equipped for that war. The troops had just gotten rid of World War I helmets and were still using a lot of World War I equipment. The radios were from the early 1930s. Still, the division fought well, a lot better than the history books give them credit for.

General Jack Vessey (left) with Bob Brandt, at the dedication of Minnesota's World War II Memorial in St. Paul, Minnesota, 9 June 2007. Brandt, VUMS Minnesota State Commander, was instrumental in recruiting General Vessey as a VUMS member.

As first sergeant, Jack's main job was to keep the troops alive, to make sure they got fed, and to make sure the company got replacements. He would get his tail chewed out for everything that didn't work. A major victory for the Allied forces took place on Hill 609, but he recalls another side of the battle that didn't make the history books. The night before, atabrine tablets (for the prevention of malaria) had been handed out to the troops. The next day there were more guys sick from the atabrine than there were casualties from the Germans.

After the Germans were out of Africa, the 34th Division was encamped along the Mediterranean Sea near Bizerte. However, the Germans still had a lot of air power and the area would be bombed frequently. Jack was wounded during one attack during the summer of 1943.

It didn't take long for him to get over his wound, but he had another medical problem, a cyst. The Army doctors were not at their best at helping it heal. He got word that if he didn't get back to his outfit soon, he'd lose his job, so he went AWOL from the convalescent camp. For a short while he was listed as a deserter.

In late summer 1943, the division packed up and headed for Italy, landing at Salerno on DDay+5 or so. The advance of the Americans up the boot of Italy was just a long series of tough battles. The Voltumo River was crossed and recrossed - many times. By Christmas 1943 they had reached Cassino, and early the next year the 34th Division was the first to enter that famous monastery town.

The division was pulled back from Cassino to join the invasion at Anzio. One day Jack was called back to headquarters, along with four or five other guys. Jack was told that they wanted to make him a second lieutenant. There didn't seem any way that he could say no. By this time the division had been in war a couple of years, and Jack knew that second lieutenants were pretty much cannon fodder. He thought it was okay to be an officer, but it wasn't like he'd just won the lottery. He felt proud that they thought he was a reasonably decent soldier.

That first winter in Italy, clothing for the troops was just marginal for the weather. There was snow and mud, wet and cold, all the time. The German shelling and bombing added to that discomfort. It was difficult to move artillery. Bulldozers had to be used to pull things around. The Germans never left a bridge behind, so rivers had to be forged.

With his commission came the job of being a forward artillery observer. At one point, both the battery commander and executive officer were wounded at the same time. For all intents and purposes, Jack was the battery commander for some time.

Jack stayed a second lieutenant until long after the war was over. The 34th Division had an unwritten rule that no second lieutenants with battlefield commissions would be promoted. He went from being the youngest first sergeant in the Army to perhaps the most senior second lieutenant.

As the war ended in Italy, his number came up to be rotated home. He was assigned the job of escorting an entire shipload of German prisoners to Naples, where he was to catch a flight home. To say there was chaos would be an understatement. There was a long line waiting for an airplane to go home. Somehow, Jack got himself designated as a courier and eventually used that precedence to get an airplane seat.

After World War II, he didn't know whether to stay in or get out, but he decided to stay in. He had his records corrected to his true age. In April 1946, he was finally promoted to first lieutenant. He attended Army and civilian schools and rose through the ranks. As a lieutenant colonel in Vietnam, he led a 300-man artillery unit against an all-out attack by 2,000 enemy troops. At one point, he picked up a grenade launcher and knocked out three enemy rocket launchers. He was awarded the Distinguished Service Cross for that action.

By 1970, he was a brigadier general and had assignments in Thailand and Laos. He received a fourth star in 1976 and commanded the forces in Korea. He became Army Vice Chief of Staff in 1979. In 1982, President Ronald Reagan named Jack Vessey, the former 16-year-old private, to be the Chainnan of the Joint Chiefs of Staff for a four-year term. General Vessey retired from the Army in 1985 after 46 years of active duty. At age 85, he still remembers that 16-year-old kid that reported for duty with the 34th Infantry Division in 1939.

General Vessey's service to his country was not over after his retirement from the Army. He led the U. S. Mission to Vietnam that resulted in the return of remains of many servicemen who had been listed as missing in action. He still serves on several national commissions, including those looking at the national security of the United States. He was awarded the Medal of Freedom by President George H. W. Bush in 1992. Jack and his wife Avis live in Garrison, Minnesota. They have two sons and a daughter.

# Dan Warden

*Age 14 - Army National Guard*

He wasn't gone all that long – just 19 months. Not long for a young kid. He was 17 years and four months of age when he came home from Korea in January of 1954. The thing is, people his age should never have to go through what Dan Warden went through. It is painful to him to even think about it. He says "I do not want to talk about it. Better to forget". But he can't. Respecting that, I have cleared this story with him, and I have left out of this account most of what still hurts him.

His age as stated above is not a missprint. In fact, he was not yet 15 when he joined the California Army National Guard.

Daniel D. Warden was born on September 25, 1936 in Shafter, CA. Shafter is a small farming community in the southern part of the central valley about 18 miles northwest of Bakersfield. Not much goes on there. But, in his pre-teen years he spent a lot of time in the fields with his mother and brother picking cotton. It was hard work dragging a 25 ft. harvesting bag and filling it with cotton in the hot interior valley climate of California. Other than that, Dan took in the football games of the Shafter Generals. That was pretty much it.

He moved with his family to Anaheim, and attended John C. Fremont Junior High. He went on to Anaheim High School and worked as an usher at the Fox Theater. He did see the school's Colonists play football, but he preferred working to earn his own money.

All of his friends were in the Army National Guard so he decided he would join to be with them. They were older, but he had to show them he could also do the job. It was February 1952. He had to lie about his age to get into the Guard. He was 14 1/2. He claimed he was 17, but was told that was too young to join without written parental consent. So he got the necessary papers and said he would be back after he got them Signed. He was soon back. The recruiter expressed surprise at the quick action. He told the recruiter his mother worked just a few doors down the street. Voila! He Was in! (it is worth noting that Dan forged his mother's name).

Anyway, he was now in the California Army National Guard, 40th Infantry Division (Mechanized), nicknamed the Sunshine (or Sunburst) Division. He took his basic training at Fort Hunter Liggett. After basic, and together with one battalion added from the Kansas National Guard, they were soon headed for the Far East as the United States 40th Infantry Division. Dan was 15 years old when he arrived in Korea in June of 1952.

He was in no way prepared to handle the horror he encountered there.

Back home life was simple, safe and without much in the way of demands. There were no drugs, —- no violence on TV (in fact no TV), —- no road rage, —- no street gangs, —- no drive-by shootings, —- no automatic assault rifles, —- no counterculture. For the most part everyone just minded his or her own business. Korea on the other hand, was like a worst possible nightmare, an experience he couldn't begin to have imagined.

Everywhere there was destruction. The civilian population was broken: thousands were killed, wounded, orphaned or left without homes. Businesses were ripped apart by bombing and gunfire. Roads, railroads and communication lines were destroyed. It was June of 1952 and Dan Warden was just a kid of 15.

At Panmunjom, talks between the Communists and the UN had been underway since early in the year of 1951. On July 26 of that year the two delegations finally agreed on a 5-item agenda to bring about a cease fire and an armistice. Chief among the agenda items were: fixing a military demarcation line and a demilitarized zone; concrete arrangements for a cease-fire; and arrangements for exchanging POWs. But negotiations were hardly carried out in good faith. Both sides mistrusted the other; suspicious of each other's motifs, and not inclined to give away any hard-won advantage. As a consequence, the talks drug on and on —- for two years. Meanwhile, fierce fighting continued all along the front.

But the overall character of the war changed. Both sides focused on improving their battle line positions. Rather than launching forward with major new assaults, time was spent on strengthening what they already had with fortifications and limited-objective attacks aimed at securing positions of advantage along the overall front. These battles resulted in the loss of thousands of lives and went into the record books as Bloody Ridge, Heartbreak Ridge, Punchbowl, Sandbag Castle, Iron Triangle, Pork Chop Hill, Bunker Hill, Old Baldy among many others.

The 40th Sunshine Infantry Division deployed to Korea in January 1952. It moved north in February to relieve the 24th Infantry Division on the battle line in the Eighth Army's X Corps sector in the rugged mountain terrain of the eastern part of the country. Dan arrived there in June. The 40th participated in the battles of Sandbag Castle and the Punchbowl. The Division suffered 1,180 casualties. Including 311 killed in action 4nd 47 more who later died from their battle wounds. Dan Warden was just a kid of 15.

Corporal Warden's Korean tour of duty ended in January 1954. He was just a kid – 17 years of age.

Dan was honorably discharged from the service, an old man at the age of 19. His most memorable experience? "I made it home in one piece."

Dan went on to spend a hitch in the navy. From there he served three years with the City of Portland Police Department. Then he drove a truck for- 32 years. Retired now, he lives in the quiet and peaceful little town of Selma, Oregon in Josephine County.

We are grateful, Dan, for what you have done for us and our great nation and for your service in the cause of freedom and world peace. You are a true hero. Your story will help to inform Americans of this and future generations of the sacrifices made by tens of thousands of our troops to protect and preserve the freedom we are so blessed with and the liberties we cherish and enjoy in our present time.

Corporal Daniel D. Warden, we, your fellow Veterans of The Korean War, salute you.

# George R. Wise, LM2490

*Age 16 - North Carolina National Guard*

I joined the North Carolina Army National Guard in Charlotte, North Carolina, on 25 October 1939. I was only 16 years old. My father had passed away in 1938, leaving my mother, my sister and me without much money. I had to scrape for what I had. The National Guard was paying $2.00 per drill and they drilled twice a week. The reason I joined the Guard was to get $4.00 per week. I liked the Army and made private first class in about six months.

President Roosevelt mobilized the Guard for one year of training. My unit, Company F, 120th Infantry Regiment, reported to Fort Jackson, South Carolina. Soon after reporting to Fort Jackson, I was promoted to corporal and made squad leader in a rifle platoon. I served in this capacity for a month or two, then I was assigned to the 30th Division training cadre. We trained the first group of draftees that was inducted into the Army in February 1941. By this time I had turned 17; my birthday was on 24 February 1941.

I served in the training cadre until about 15 May 1941. I heard of a vacancy for a company clerk in an ambulance company. I was tired of walking in the infantry and I was a good typist, so I applied for the job. I was accepted and was transferred to Company F, 105th Medical Regiment of the 30th Division. I stayed in that unit until 15 September 1941 when my granddaddy, who was a judge in North Carolina, got me out of the Army.

I went home and went back to high school.

Times were pretty hard at that time, but I carried papers and worked after school, so we made out okay. When the Japanese bombed Pearl Harbor I was recalled to active duty. I went back to the same ambulance company that I had left in September.

I was still only 17 when I was recalled. After I turned 18 in February 1942, I didn't think anymore about my age until I was in officer candidate's school (OCS). I was ordered to the Medical Administration OCS at Camp Barkley, Texas, in April 1942. Three weeks before graduation I was told that I had to have a birth certificate. I went to my captain and told him that I was only 18 and that I had lied about my age when I enlisted in 1939. The captain told me that I would have to go before the general (a brigadier general). I was scared to death as I had never seen a general before and I thought that they would court-martial me.

I reported to the general in my best military posture and told him what I had done. He was very nice and told me that I had done a good job in the OCS, but that he could not commission me. He told me about a new unit that was being organized in Fort Hamilton, New York, that needed men with my experience. He said if I wanted to go there that he would make me a sergeant and send me there.

Well, I took the assignment and reported to Fort Hamilton on 15 July 1942. This unit, the 2nd Group Regulating Stations, was to depart for overseas immediately. However, they split the unit into two groups and my group stayed at Fort Hamilton where we guarded a military prison until 4 March 1943, when we departed for overseas. We landed at Oran, North Africa, and established road blocks and check points all over North Africa. When the Americans and British ran the Germans out of North Africa, our check points were demolished. My unit then performed various jobs in and around the Port of Oran. I was assigned the job as port dispatcher. I dispatched trucks to shipside and unloaded the cargo. I had been promoted to staff sergeant and held this job until my unit was sent to Italy.

We landed in Naples, Italy, on or about 15 May 1944. My unit performed different missions until we were moved off the line to train in the Tyrenian Sea for an invasion. We did not know where the invasion would take place and didn't learn of our destination until we boarded the ships in Naples Harbor. We were told that we were going to invade Southern France near St. Raphael, France. We were part of the 7th Army. We were to join up with the forces that invaded Normandy on 6 June 1944.

We went ashore in a landing craft on D-Day.

St. Raphael is on the coast between Nice and Marseilles. We advanced up the eastern border of France and joined up with the 1 SI Army at Luneville, France. I was promoted to technical sergeant while at Luneville.

We established convoy check points and made reconnaissance for convoy routes. We were continually on the move until we reached Saarburg. When the Battle of the Bulge started, we moved back to Luneville, France, and reestablished the 7th Army Headquarters.

On 24 February 1945, I received a battlefield commission as a 2nd lieutenant in the Transportation Corps. I performed several jobs in reconnaissance while at 7th Army Headquarters. Later, I was reassigned to the 4382nd Transport Truck Company as a platoon leader. We hauled supplies to the regimental supply points all over France and Germany until the war was over.

After the war was over, I was transferred to battalion headquarters as the battalion supply officer and stayed in that position in the Army of Occupation in Germany until I was eligible to go home. We departed Antwerp, Belgium, during the first part of October 1945, and arrived in the States on 19 October 1945.

I was discharged at Fort Bragg, North Carolina, on 11 November 1945. I remained in the reserve and in November 1951, I joined the South Carolina National Guard. I stayed in the Guard until August 1979. I retired as a brigadier general. My age was not corrected until I received my commission. I am 84 years old now and am enjoying my retirement. George Wise lives in Columbia, South Carolina.

# John Yingling

*Age 16 - United States Air Force*

He carries himself with a pilot's grace and confidence at 83 years old. And he recounts, with the astute memory of a disciplined mind, the major events of a life fully lived: He is a hero who earned the Distinguished Flying Cross. And though his heroism was little-hailed in its time, Maj. John Yingling has in recent years captured the fascination of many who long to hear his story.

He entered the U.S. Navy, like many young warriors, with crossed fingers and a white lie about his tender age - just 16 in 1941. Like his father, he felt called to do battle for his country. But where his father's dreams of service during WWI had been dashed (before he finished Officer Candidate School, the war had ended), 16-year-old John enlisted in the Navy as an aerial gunner.

After the war, he returned home and attended the University of Connecticut on the GI Bill, amazed at his good fortune: the government gave him $60 a month and the ROTC kicked in another $20. When he graduated, he was about to embark on a lucrative marketing career, but his plans were impeded by a call to serve in the Korean Conflict.

He was again called to flight service - this time as an Air Force officer piloting a P-51 Mustang, "Miss Dee," named for his first wife. It was a day he retells with clarity.

Early one morning in May, 1952, a dozen Mustangs headed deep into enemy territory to bomb the rail lines below the Yalu River between North Korea and China. After completing a successful mission, he was stunned to find his plane heading for a mid-air strike with MIG-15. He pulled his sights and fired on the enemy plane. His wingman watched as the MIG pilot bailed out - confirming that he had hit his mark.

It was the only known incident in history of a P-51 Mustang fighter pilot shooting down a MIG-15 during the Korean Conflict.

For reasons that fade into the unknown, Maj. Yingling was neither credited nor decorated for his mission until six months later. He speculates that the military and political climate of the day intervened - including a focus on the sophisticated tecnology of the new F-86 as the tool to wipe out the enemy's powerful MIG. He says: "No one wished to acknowledge that a little prop plane could do what the F-86s were touted for."

The official downplaying of his unforgettable mission did not deter his desire to serve. Maj Yingling finished out his 20-year career in missile operations safety - a field that evolved into the private sector where he worked as manager of a safety program at Kennedy Space Center for McDonnell Douglas. "Throughout my entire working life, I got up every day to an exciting job," he recalls.

Maj. Yingling and his wife Corinne moved to AFV in 1998 and long lent their singing talents to one of the Village's most popular groups, the Village Voices. He also has been involved in the Residents' Council where he served a stint as chair. When his wife's health declined, he retired from many of his activities and became her loving caretaker. Although he describes his care taking abilities as "amateur," he admirably performs his duties the way he approaches life - with skill and diligence.

Recently, Corinne went into the AFV Health Care Center and for Maj. Yingling it has been a painful situation fraught with mixed emotions. As much as he misses his beloved wife, he knows she is getting excellent care. "It's hard to find people who have the compassion I see there every day, and very soothing for families to see that level of kindness." He says.

His memories are splendid of a life well lived. Two gratifying careers. Two wonderful wives (following Dee's death in 1980, he married Corinne). Four children, 14 grandchildren and four great-grandchildren. Ten years of delightful RV travel during which he and Corinne criss-crossed the country five times. And a great retirement at AFV, a decision that reminds them, he says, "of our active military days. You would come to a new base and make a circle of friends that became like family. It has been a welcome relief to return to a military environment.